Laminar Drag Reduction

Authored By

Keizo Watanabe
Tokyo Metropolitan University
Japan

CONTENTS

FOREWORD

Drag reduction is an area of research and development in fluid mechanics, where the energy efficiency of fluid transport systems can be improved by controlling fluid flows. For example, one can reduce the fuel consumption of aircraft, ships, trains and motorcars by applying different drag reduction technologies. Here, using less fuel does not only imply less emissions of harmful gases to environment, but also lead to a reduced noise level, contributing much to an enrichment of the quality of life. The Advisory Council for Aviation Research and Innovation in Europe (ACARE) has recently published the EU's vision for future aviation, where CO_2 and NOx emissions should be reduced by 75% and 90%, respectively by the year 2050, accompanied by a noise reduction of 65%. A new EU framework programme for research will be launched later this year with a budget of € 80 billion in order to achieve an economic growth of a sustainable society, where drag reduction technologies will play an important role in achieving breakthrough innovation.

Drag reduction can be obtained by controlling either turbulent or laminar flows. Turbulent drag reduction has been an active area of research, where turbulence can be supressed by flow surface modifications such as riblets, or drag-reducing additives such as long-chain polymers and surfactants. The latter is the drag reduction technique that the author of this book has extensively investigated over past years. The laminar drag reduction, on the other hand, is a relatively new areas of research, where the laminar flow can be controlled by microscopic surface modification, allowing the flow to slip over the wall. Here again, the author has contributed in establishing the mechanism of drag reduction by hydrophobic surface.

The majority of this book's contents comes from the author's own research. The first chapter gives an overview of drag reduction methodology and an introduction to the concept of fluid slip. This is followed by experimental and analytical results of flows through circular pipes and ducts with hydrophobic surface, where both Newtonian and non-Newtonian flows are discussed. Throughout these chapters, the mechanism of laminar drag reduction can be explained as a result of slip flows over microscopic patterns created by hydrophobic surface, where the mean velocity profile gradient is reduced. Flows between coaxial cylinders and over a rotating disk are discussed next, where changes in flow patterns over hydrophobic surface are described. The measured torque on rotating cylinders and disks confirmed that the laminar drag reduction can also be obtained over rotational components. The final two chapters of the book deal with flows over circular cylinders and spheres with hydrophobic surface, where experimental data are compared with numerical simulation results at low Reynolds numbers.

This book serves as a comprehensive guide to the latest information and understanding of laminar flow control using hydrophobic surface. It is highly recommended to postgraduate students, academics and researchers as well as to design and practicing engineers.

Kwing-So Choi
University of Nottingham
UK

PREFACE

Hydraulic transportation systems with a pumping operation are frequency used in petroleum pipelines or the process lines of many industrial plants. Thus, approaches for reducing the pumping power promote energy savings. For example, we highlight the application of drag reduction phenomena as one such procedure that can reduce the transportation power.

As well known, in 1883 O. Reynolds described the flow behavior in a pipe, namely, the real fluid flow regimes which are classified as laminar or turbulent on the basis of the flow structure. Flow structure in the turbulent regime is characterized by random motions resulting from the turbulence, which is a dominant factor for the drag. Therefore, if the flow is turbulent, modification of the turbulence becomes the target for achieving a reduction of the drag. Since Toms' effect was reported in 1943, many researchers have investigated the turbulent drag reduction that is achieved using numerous drag reducing additives, including high molecular weight polymers, micro fibers or particles, surfactants and bio-polymers *etc*. Currently, polymer or surfactant solutions are applied in many pipeline systems.

On the other hand, in the laminar regime, the flow structure is characterized by smooth motion in the layers with no turbulence. Thus, it is necessary to establish a new concept within the turbulence modification procedure in order to obtain laminar drag reduction. In general, there is no slip at the boundary for real fluids; the fluid in direct contact with the solid boundary has the same velocity as the boundary itself. This is an experimental fact based on numerous observations of fluid behavior. If fluid slip occurs at the solid boundary, the drag or loss will be reduced compared to that of the case of no slip; laminar drag reduction is achieved in this scenario. The development of a hydrophobic material makes it possible to cause a relatively large fluid slip at the solid wall. Although the fluid slip is not significant for achieving drag reduction in turbulent regime, using a fluid slip we can obtain drag reduction in a laminar flow.

This book was written to meet for a discussion of laminar drag reduction utilizing the fluid slip of Newtonian fluids at a highly water repellent wall. The author hopes that this book will serve the need they see and be useful researcher and engineering at universities and for the practical engineer on new drag reduction phenomena related the interaction between liquid and the hydrophobic wall.

Many sources for the experimental results in this book have been drawn from papers that were produced at the Fluid Engineering Laboratory in the Faculty of Engineering at Tokyo Metropolitan University. I am indebted to a great many teachers and students who have assisted me during my work at the university. This book would never have been written without their cooperation.

ACKNOWLEDGEMENTS

The author acknowledges gratefully the assistance and cooperation of Dr. S. Ogata, Associate Professor at Graduate School of Science and Engineering of Tokyo Metropolitan University.

CONFLICT OF INTEREST

The author confirms that this ebook contents have no conflict of interest.

Keizo Watanabe
Tokyo Metropolitan University
Japan
E-mail: keizo@tmu.ac.jp

CHAPTER 1

Introduction

Abstract: Fluid drag or friction depends on the physical characteristics of the fluid and solid surface. It is very important to control fluid flow and achieve drag reduction. To this end, it is necessary to understand the phenomenon of drag reduction from the viewpoint of energy saving in hydraulic transportation systems. Studies on flow drag have greatly contributed to the development of hydraulics and fluid mechanics. Although we can compute the drag of a blunt body or pressure loss of a channel in various sorts of fluid flow by using the systematized knowledge gained thus far, still some unexplained phenomena have occurred in actual flow-field. Thus, it can be considered that there are two research directions; one concerns energy saving in practical applications; and the other involves the clarification of the phenomenon due to changes in the physical characteristics of fluids and surfaces during experiments. With regard to energy saving, it is possible to design and construct apparatus for simulating drag reduction by using experimental data. In this section, a drag reduction technique employing either drag reducing additives or drag reducing walls is explained. In addition, some existing experimental data on the characteristics of drag reducing walls used in laminar drag reduction are summarized.

Keywords: Drag reduction, passive control, drag reduction additives, drag reduction walls, fluid slip, Navier's hypothesis, hydrophobic wall, highly water repellent wall, wettability, contact angle.

1. CLASSIFICATION OF DRAG REDUCTION

In general, drag reduction can be achieved by the use of an energy-supply (referred to as active control) or in the absence of an external energy supply (referred to as passive control). In addition, it is necessary to investigate the fluid or surface physical properties to achieve drag reduction in a flow system because fluid drag or friction is closely related to the physical properties of the fluid and the solid surface in contact. In the passive control, the phenomenon of fluid drag is observed, and it is controlled using drag reducing additives and surfaces. Table **1** lists some methods of the passive control for drag reduction.

Drag reduction by high-molecular-weight polymer additives was discovered by B. A. Toms [1] in 1948, so the phenomenon is called Toms effect. He experimentally established that in a certain concentration range, the pressure drop in a turbulent pipe flow of a solution of polymethyl methacrylate in monochlorobenzene is less than that in the solvent. After this, this phenomenon has been studied experimentally and analytically for both internal and external flow systems. The first commercial use of polymeric drag reducing additives to increase the flow rate in a crude oil pipeline was in 1979 for the trans-Alaska Pipeline System [2]. However, the practical application of this additive is limited by the fact that mechanical shearing causes degradation of the macromolecular chain.

Drag reduction occurs in surfactant solutions whenever the surfactant molecules form rod-like micelles. This principle has been recently applied to the heating and cooling pipelines of hydraulic transportation systems, where no degradation has been observed. Although large drag reduction is achieved in liquid flow by adding high-molecular-weight polymer or surfactant additives, solution degradation or leakage from a pipeline may occur. Therefore, the practical application of surfactant additives is limited to closed pipeline systems. Yeast and rice malt fungus, which produce a polysaccharide, are used as drag reducing biopolymers. Xanthan gum and guar gum are other examples of such biopolymer. These natural polymers are expected to attract attention owing to their environment friendly characteristics.

Some types of wall surfaces, *e.g.*, surfaces with microscopic roughness, such as shark skin, have been experimentally shown to be effective in turbulent drag reduction. Application of drag reducing walls has advantages of low production cost and of maintenance, in general. However, major disadvantage of this technique is low drag reduction as compared to that of drag reducing additives. Riblets are small surface protrusions aligned with the direction of flow, which confer anisotropic roughness to a surface. It is known that longitudinal ribs fabricated on a flat surface reduce the turbulent skin friction drag by about 8% by limiting range of turbulent flow. The merits of riblets lie in the fact that they can be used in both aero and liquid flow systems, which has been confirmed by performing flight and shipping tests using riblets films, although the maintenance costs are very high as these can be blocked by the dust particles very easily.

All methods listed in Table **1**, except the one employing a hydrophobic wall, affect the range of turbulent flow, and this often results in turbulent drag reduction. Turbulent drag reduction involves changing the velocity profile of the flow and modifying its turbulence as discussed above. In contrast, the hydrophobic wall causes laminar drag reduction by creating a flow with no turbulence in the laminar region as shown by O. Reynolds' through flow visualization experiments in 1883. Therefore, the concept of turbulence modification cannot be applied to laminar drag reduction. In laminar flow of Newtonian fluids, the wall shear stress is simply the product of viscosity and shear rate at the wall as shown by Newton's law of viscosity. If fluid slip occurs on the wall of a flow channel in a pressure-driven flow system, the flow rate of the channel is clearly found to increase as compared to that in a no-slip condition. In other words, drag redaction occurs in this case. Presently, hydrophobic walls are used to achieve laminar drag reduction by inducing fluid slip in a liquid flow. The fluid slip discussed here is an apparent fluid slip, and its occurrence is restricted to the case where the Reynolds number of a liquid flowing along a hydrophobic wall is low.

Table 1: Classification of drag reduction

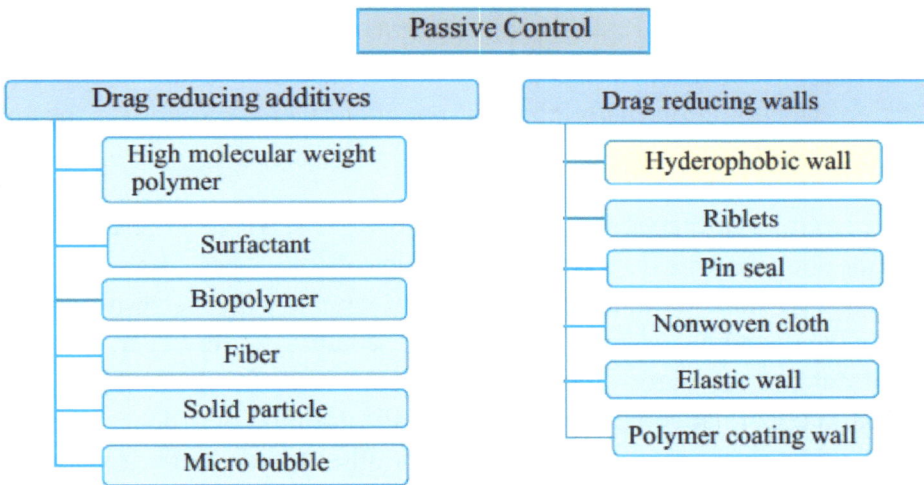

```
                         ┌─────────────────────────┐
                         │    Passive Control      │
                         └─────────────────────────┘

┌─────────────────────────────┐       ┌─────────────────────────────┐
│   Drag reducing additives   │       │    Drag reducing walls      │
└─────────────────────────────┘       └─────────────────────────────┘
    ┌─────────────────────────┐           ┌─────────────────────────┐
    │ High molecular weight   │           │   Hyderophobic wall     │
    │ polymer                 │           └─────────────────────────┘
    └─────────────────────────┘           ┌─────────────────────────┐
    ┌─────────────────────────┐           │        Riblets          │
    │       Surfactant        │           └─────────────────────────┘
    └─────────────────────────┘           ┌─────────────────────────┐
    ┌─────────────────────────┐           │       Pin seal          │
    │       Biopolymer        │           └─────────────────────────┘
    └─────────────────────────┘           ┌─────────────────────────┐
    ┌─────────────────────────┐           │    Nonwoven cloth       │
    │         Fiber           │           └─────────────────────────┘
    └─────────────────────────┘           ┌─────────────────────────┐
    ┌─────────────────────────┐           │     Elastic wall        │
    │     Solid particle      │           └─────────────────────────┘
    └─────────────────────────┘           ┌─────────────────────────┐
    ┌─────────────────────────┐           │  Polymer coating wall   │
    │      Micro bubble       │           └─────────────────────────┘
    └─────────────────────────┘
```

2. BASIC CONCEPT OF FLUID SLIP

In 1823, L. Navier [3] formulated a hypothesis for fluid slip by considering the difference between experimental data of flow rate in previous works and the analytical result of applying the Navier-Stokes equations to fully developed laminar flow in a pipe. He hypothesized fluid slip from the fact that an experimental flow rate was greater than that given by the exact solution. The hypothesis stated that fluid slip occurs in fluid and the slip velocity increases in proportion to the velocity gradient, as follows:

$$u_s = \frac{\mu}{\beta}\left|\frac{\partial u}{\partial y}\right| \qquad\qquad (1\text{-}1)$$

where, u and β are fluid velocity and sliding constant, respectively. u_s is the slip velocity and the equation gives the slip velocity at the wall if the wall shear rate is derived.

In light of this hypothesis, we can conclude that the flow rate does not increase as a result of fluid, which was confirmed by comparing the experimental data. Experimental data of laminar flow agree with the exact solutions of the Hagen-Poiseuille formula calculated under no-slip conditions. Thus, Eq. (1-1) is not a hypothesis drawn from the actual phenomenon; rather it has been used frequently to describe slip boundary conditions, because it is a very simple equation. However, difficulties lie in the measurement and determination of the value of the sliding coefficient β for each flow condition at present because slip phenomenon causes friction depending on many physical parameters of the fluid and the solid surface.

On the other hand, regarding the phenomenon of fluid slip, the experiment leads us to conclude that viscous fluid does not slip at solid surfaces as shown by the fact that many experimental data are in exact agreement with the result given in the equation, analyzed by a no slip boundary condition. However, based on experimental data, it has been reported in the recent years [4-7] that fluid slip of Newtonian fluid occurs at some kind of solid wall of a pipe or a duct in pressure driving flow although its value is very small. In other words, experimental flow rate increases in comparison with that of the exact solution. More recent investigations [8] were reported for fluid slip of large flow system size.

Fig. 1 shows the experimental results of previous studies on the relationship between flow system size, L and Reynolds number, Re for fluid slip. Slip flow is confirmed by measuring the pressure losses or velocity profiles of a pipe or a duct. $L=10^{-2}$ m for the maximum flow system size. This is called a hydrophobic wall or highly water-repellent-wall, and nano-bubbles have been observed in some experimental micro-channels. As the phase of these bubbles may provide a zero shear stress boundary condition, an apparent fluid slip occurs in the case of liquid fluid flow. However, the mechanism of apparent fluid slip cannot be explained only in terms of nano-bubbles for large flow system. In the case of fluid slip in a large flow system, it is important to point out the presence of a relatively modest gas interface between the wall and the liquid, whose size is larger than that of the nano bubbles. Therefore, the question arise: how can this interface exist at the wall surface under pressure driven flow? Experimental visualizations have indicated that the wall should be highly water repellent, and that should be a fractal surface with small clacks. The fluid slip occurring through this phenomenon is called apparent fluid slip.

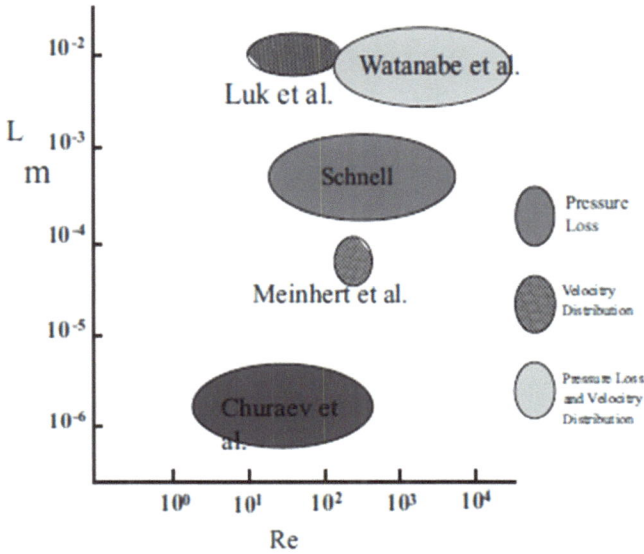

Figure 1: Flow system size of previous study.

3. PROPERTIES OF HYDROPHOBIC WALL WITH APPARENT FLUID SLIP

A droplet maintains a spherical shape because of intermolecular attraction. It becomes misshapen when it comes in contact with the solid surface, because of molecular interaction between liquid and solid. If intermolecular attractive force which acts between solid surface and a droplet increases, the shape of the droplet becomes flat. Conversely, the droplet becomes spherical when the force is very small. The first is an example of a hydrophilic wall and the second is a hydrophobic wall. In general, we can obtain highly hydrophobic properties by making a microscopic patterned indented surface or selecting a material for the solid surface with low surface free energy.

We can often observe a sub globular droplet on a lotus leaf, as shown in Fig. **2**. It is well known that a lotus leaf and an aroid leaf are highly repellent surfaces; they are examples of a hydrophobic wall. These surfaces appear fluffy because of the glandular hair of the plant fiber, a kind of rough surface with a fractal structure, which can be seen in a microscope. The microscopic bumpy surface increases the actual surface area, and the contact angle of the droplet increases as shown by Cassie's equation [9]. The contact angle is a measure for the wettability of the surface; it is the angle of the line tangent to the droplet surface and the solid surface. The angle is $180°$ in the case of a spherical droplet, which means it is a perfect hydrophobic wall. It is shown that a smooth surface of polytetrafluoethylene (PTFE) provides the largest contact angle surface, a material having no fractal quality. The angle measures about $110°$. Thus, it is necessary that we make a fractal surface with many fine grooves in order to obtain a wall with low wettability (in other words a highly water-repellent wall).

Figure 2: Droplet on a lotus leaf.

There are physical and chemical methods to produce a fractal surface. Chemical photo etching is one simple methods to make various sorts of fractal configurations on a surface by artificial means. Fig. **3a** shows an example of a silicon surface with a random combination pattern, obtained by the etching; Fig. **3b** is a droplet on a surface. The detailed view is shown on the right side of Fig. **3a**. The Y shape is etched and uniformly spaced. The Y is 5μm wide and 10μm deep. The measurement showed that contact angle of the surface with the droplet is larger than what it would be for a smooth surface with no etching. The chemical photo etching technique is an easy-to-use method for obtaining a highly water repellent wall with large contact angle. However, it is very difficult to etch on a circular pipe wall. For this case, coating or splaying with PTFE liquid produces the highly water repellent wall desired, because it is a material with low surface free energy.

Fig. **4a** shows a microphotograph of a PTFE coated surface obtained by using a microscope and scanning electron microscope (SEM). As shown in the figure, there are many fine grooves on the surface, which raise water repellency. The basic material substance of the highly water-repellent coating is a fluorine alkane-modified acrylic resin with added hydrophobic silica, which was left overnight in air after coating the wall. Fig. **4b** shows the shape of a droplet of water on the surface. The contact angle with the wall is about 150°. In general, the largest contact angles recorded for smooth surface are 112°-115°. Thus, we need a useful method for producing a surface with a contact angle of 120° or larger, since we must not only reduce free surface energy but also change surface morphology.

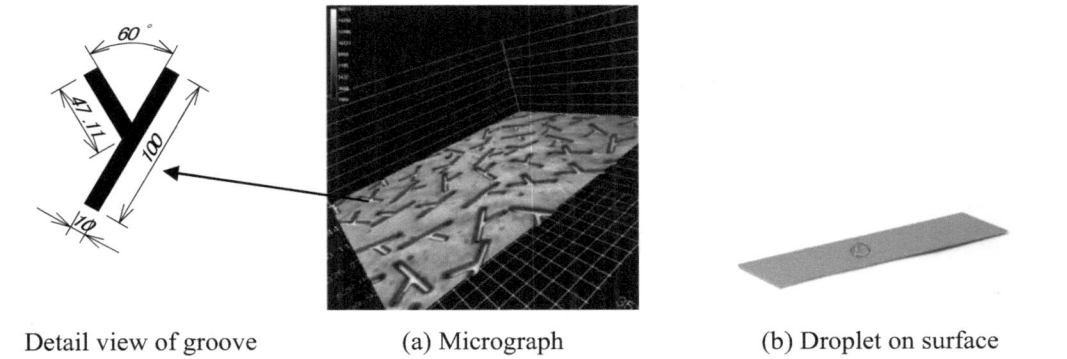

Detail view of groove   (a) Micrograph   (b) Droplet on surface

Figure 3: Silicon plate surface made by etching.

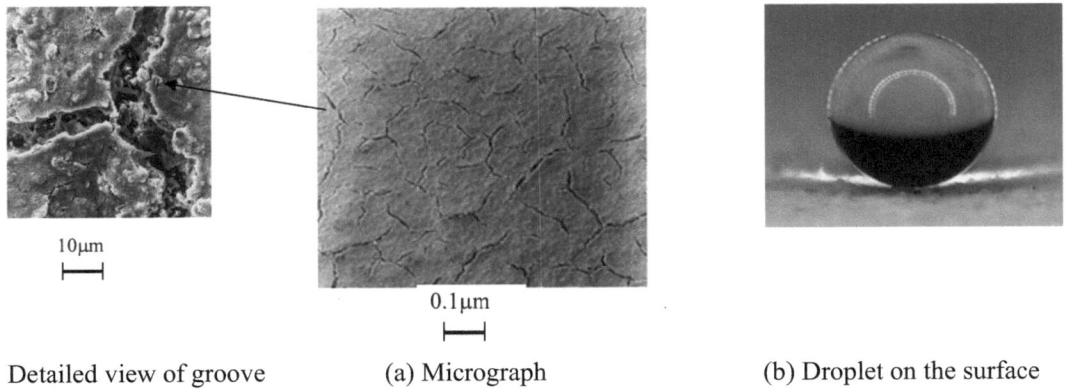

Detailed view of groove   (a) Micrograph   (b) Droplet on the surface

Figure 4: Hydrophobic surface coated with PTFE paint.

The contact angle is an intuitive evaluation index of surface wettability in the static condition. Although the water repellency of the surface can be evaluated viscerally from this angle, it cannot be directly used to analyze the apparent slip of the liquid. The problem remains in the correlations between the apparent slip velocity and the physical constant of the wall surface or liquid of the phenomenological model. If it was feasible to correlate the association between these physical constants and the sliding coefficient in Eq. (1-1), we may be able to analytically evaluate or predict the apparent fluid slip velocity. On the other hand, flaky paint at the flow channel wall is a highly water-repellent material because it has very low interfacial free energy. From the practical viewpoint, the hydrophobicity and adhesion in the physical properties of paint are contradictory in nature. Thus, an important technical problem exists in providing an attachment for the hydrophobic material to increase an adhesion property of the material without degradation, in order to maintain the pipe's physical characteristics.

REFERENCES

[1]   Toms B. A. (1948) "Observation on the flow of liner polymer solutions through straight tubes at large Reynolds numbers", Proc. Int. Rheologic. Cong., **2**, 135-141.

[2]   Burger, E. D. (1982) "Flow Increase in the Trans Alaska Pipeline Through Use of a Polymeric Drag-Reducing Additives", J. of Petrol. Tech., **34-2,** 377-386.

[3]   Navier, C. M. L. H. (1823) "Mémoire sur les lois du movement des fluids", Mémoires de l'Académie Royale des Sciences de l'Institut de France, **6,** 389-440.

[4]   Schnell, E. (1956) "Slippage of water over nonwettable surface",. Appl. Phys., **27**, 1149-1152.

[5]   Churaev, N. V., Sobolev, V. D. & Somov, A. N. (1984) "Slippage of liquids over lyophobic solid surface", J. Colloid. Interface Sci. **97**, 574-581.

[6]   Tretheway, D. C., Meinhert, C. D. (2002) "Apparent fluid slip at hydrophobic microchemical walls", Phys. Fluids **14**, L9.

[7]   Luck, S., Mutharasan, R., Apelian, D. (1987)"Experimental observations of wall slip: tube and packed bed flow", Ind. Eng. Chem. Res., **26** (8), 1609-1616

[8]   Watanabe, K., Yanuar and Udagawa, Y., (1999) "Drag reduction of Newtonian fluid in a circular pipe with a highly water-repellent wall", J. Fluid Mech., **381**, 225-238.

[9]   Cassie, A.B.D., Baxter, S. (1944) "Wettability of porous surface", Trans. Faraday Soc., **40**, 546-551.

CHAPTER 2

Newtonian Fluids Flow in a Circular Pipe

Abstract: Laminar drag reduction, which occurs through apparent fluid slip, was shown for Newtonian liquids flow in a pipeline system with a highly water-repellent wall pipe by measuring the pressure drop and the velocity profile. The same hydrophobic pipe was also used in experiment for a circular pipe flow, shown in Fig. **4** in Chapter 1. It is 14% in the drag reduction ratio for 12 mm diameter pipe. The friction factor formula for a pipe with fluid slip at the wall was derived analytically using the Navier-Stokes equation and Navier's hypothesis for fluid slip of the boundary condition. The result obtained using the friction factor formula agrees well qualitatively with the experimental data. It was experimentally clarified that the relation between the slip velocity and the wall shear stress is a substantially linear relation. Because the sliding constant is given by the gradient of an approximated straight line from Navier's hypothesis, the comparisons between the experimental data and the analytical result are quantitatively enabled by substituting the value for the friction factor formula. Measurement result of the velocity profile shows the occurrence of slip velocity at the wall, and the cause of slip is discussed It can be considered that the micro bubble has no effect for the slip velocity since this flow system size is order of 10mm. Experimental result of surfactant solutions without the laminar drag reduction suggests the existence of air-liquid interface at the wall.

Keywords: Drag reduction, laminar flow, pipe flow, Newtonian fluid, pressure drop, friction factor, velocity profile, highly water-repellent wall, fluid slip, slip velocity, sliding constant, wall shear stress, degassing solutions.

1. INTRODUCTION

Circular pipe flow is one of the most basic pressure-driven-flows. It occurs in many engineering systems such as crude-oil transportation pipelines, water and sewerage and in the plumbing of industrial plants. It is widely known that the formula for the pressure drop of the fully developed flow can be obtained experimentally and analytically for Newtonian or Non-Newtonian fluids for the case of no fluid slip. Because the formula for the friction factor is applied as a criteria value to that of drag reduction, it is important to derive the exact solution of Navier-Stokes equation under fluid slip boundary conditions for equation with experimental results. First, the velocity profile of Newtonian fluids is analyzed by assuming fluid slip at the pipe wall and after analytically obtaining the friction factor. The results of the analysis will enable readers to understand the characteristics of the laminar drag reduction of a pipe flow occurring by fluid slip. The main reasons for the fluid slip observed at highly water-repellent walls are

Keizo Watanabe

discussed by considering the microscopic scale at the interface between the liquid and the wall.

2. ANALYTICAL APPROACH

Which conditions need to be satisfied by a moving fluid in contact with a solid body have for a long time remained a challenging question, as pointed out by Goldstein, [1] and the assumption of no slip is now generally accepted for practical purposes. If the flow of incompressible fluids is laminar, the velocity from the entrance of a pipe is illustrated as shown in Fig. **1**, where cylindrical coordinates r, θ, and z are used. These velocity profiles [2] υ_z in the figure are calculated using the results of Langhaar's analysis [3]. In Fig. **1**, L_z, R_0, and U_0 are the entrance length, pipe radius, and mean velocity, respectively. In general, the entrance length is required to be greater than 100 pipe diameters to obtain a fully developed flow. Sufficiently far from the pipe entrance, the flow is fully developed and is widely known as the Hagen-Poiseuillie flow. The shape of the fully developed velocity profile is a quadratic curve. Our intent is to obtain detailed information about the velocity field with fluid slip. Before this, we must first discuss the phenomenon of drag reduction quantitatively.

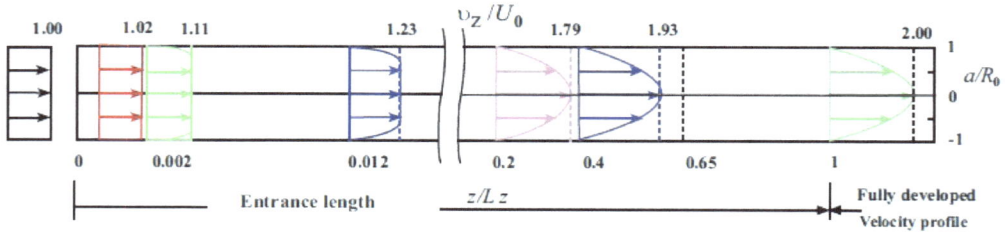

Figure 1: Velocity profiles in the entrance of a pipe.

For a fully developed flow of Newtonian fluids in a pipe, the Navier-Stokes equation reduces to

$$0 = \frac{\partial p}{\partial r} = \frac{\partial p}{\partial \theta} \tag{2-1}$$

$$0 = -\frac{\partial p}{\partial z} + \mu \left[\frac{1}{r} \frac{\partial}{\partial r} \left(r \frac{\partial \upsilon_z}{\partial r} \right) \right] \tag{2-2}$$

where, p and μ are the pressure and the viscosity, respectively. From Eq. (2-1), each term for the $r-$ and $\theta-$directions is zero, and $p = p(z)$

$$\frac{\mu}{r}\left[\frac{d}{dr}\left(r\frac{d\upsilon_z}{dr}\right)\right]=\left(\frac{dp}{dz}\right)$$ (2-3)

By integrating Eq. (2-3), and owing to the physical condition that the velocity must be finite at $r=0$:

$$\upsilon_z=\frac{r^2}{4\mu}\left(\frac{dp}{dz}\right)+C_1$$

The constant C_1 is evaluated under the boundary conditions on a fluid slip with the slip velocity u_s at the pipe wall: $r=a$, and $\upsilon_z=u_s$. The slip velocity u_s is determined using Navier's hypothesis, Eq. (1-1), from a macroscopic viewpoint:

$$\tau_w=\mu\left(-\frac{d\upsilon_z}{dr}\right)_{r=a}=\beta u_s$$ (2-4)

where τ_w and β are the shear stress at the pipe wall and the sliding constant, respectively. In the case of $\beta\to\infty$, Eq. (2-4) agrees with the no-slip condition.

Consequently, this gives

$$C_1=\left(\frac{a}{2\beta}+\frac{a^2}{4\mu}\right)\left(-\frac{dp}{dz}\right),$$

and hence,

$$\upsilon_z=\left[\frac{1}{4\mu}a^2\left(1-\frac{r^2}{a^2}\right)+\frac{a}{2\beta}\right]\left(-\frac{dp}{dz}\right).$$ (2-5)

Under the no-slip condition, $\beta\to\infty$, and Eq. (2-5) becomes the Hagen-Poiseuillie flow as shown in Fig. **1**

$$\upsilon_z=\frac{1}{4\mu}\left(-\frac{dp}{dz}\right)a^2\left(1-\frac{r^2}{a^2}\right)$$

In a fully developed flow, the pressure gradient $\left(-\dfrac{dp}{dz}\right)$ is constant. Therefore, the pressure gradient is $(-dp/dz)=(p_1-p_2)/l=\Delta p/l$.

The volume flow rate Q can be obtained as follows:

$$Q = \int_0^a 2\pi r \upsilon_z dr = \frac{\pi a^4 \Delta p}{8\mu l}\left(1 + \frac{4\mu}{a\beta}\right)$$ (2-6)

Although a friction factor is defined by the wall shear stress τ_w or the pressure loss Δp, the friction factor λ which is based on the pressure loss is used in this book. Where, it is given as $\lambda = \Delta p / \left[(l/2a)\left(\rho U_0^2/2\right)\right]$. If a friction factor is defined using the wall shear stress as $f = \left[\tau_w / \left(\rho U_0^2/2\right)\right]$, we obtain $\lambda = 4f$. Where, ρ and U_0 are fluid density and mean velocity, respectively.

The friction factor λ becomes

$$\lambda = \frac{64}{\text{Re}}\frac{1}{\left[1+\left(\dfrac{4\mu}{a\beta}\right)\right]} = \frac{64}{\text{Re}}\frac{1}{(1+4S)}.$$ (2-7)

Thus, in a fully developed laminar flow with fluid slip, the friction factor is a function of, not only the Reynolds number Re but also the non-dimensional parameter $S = (\mu/a\beta)$ including the sliding constant β given by Eq. (1-1) in Chapter 1. Because the parameter related to fluid slip is always a positive value in Eq. (2-7), it decreases compared with that of a flow under the no-slip conditions $S = 0$, which is given as $\beta \to \infty$. To summarize these results, the laminar drag reduction phenomenon can be explained by fluid slip and we can calculate the drag reduction ratio by Eq. (2-7).

3. EMPIRICAL APPROACH

3.1. Friction Factor

Experiments [4] were carried out to measure the pressure drop and the velocity profile of tap water and aqueous solution of 20-30wt% glycerin which is Newtonian fluid in the circular pipe with a hydrophobic wall shown in Fig. **4** in Chapter 1. Test pipes were approximately 6 and 12 mm in diameter. The thickness of the hydrophobic coating was less than 10 μm, and the mean value of the inlet and outlet diameters of the test pipe was measured using a micrometer. Pipes with smooth wall of the same size made of acrylic resin were tested in order to compare the experimental results under no-slip conditions. Experimental results of the pressure drop were determined using two experimental apparatuses, which were circulation-type and pressure-driven-type pipeline systems, in order to validate the evidence. In the circulation-type system, test fluids were circulated using a centrifugal pump with variable rotation speed, and a fully developed steady flow was obtained in the test section 475 mm in length. The pressure loss at the test section was measured using a pressure transducer, and the measurement was carried out two or three times to take care.

The pressure-driven-type pipeline system was operated under the conditions of degassed water to study the effect of air in tap water on the drag reduction phenomenon. Because normal tap water contains several percent of air, the tap water in the pressure tank was degassed using a vacuum pump and the test liquid was held for about six hours in the vacuum. Subsequently, test liquids were flowed out through to the test pipe by a compressor under constant pressure. Two test liquids, which were set at -300 and -600 mmHg in the vacuum tank, were used for the pressure loss measurement. Experiment for smooth wall pipe was carried out also to compare the experimental result of the highly water-repellent wall pipe.

Fig. **2** shows the experimental results for the friction factor of tap water obtained using the circulation-type pipeline system. In the figure, the solid lines indicate the exact solution for laminar flow and the Blasius formula for turbulent flow in a circular pipe. Experimental data in the smooth acrylic resin pipe agree well with these lines in laminar and turbulent flow ranges. The result of uncertainty analysis showed that the true value is believed to lie within 3.15% of the reported value for tap water in the laminar flow range in the smooth pipe.

On the other hand, the friction factors of the hydrophobic wall pipe decreases compare to these of the smooth wall pipe in laminar flow range, namely the laminar drag reduction occurs in the region. The drag reduction of the 12 mm pipe showed 14.0% in drag reduction ratio. The friction factor is inversely proportional to the Reynolds number, and the experimental data for glycerin solutions show a similar tendency. The amount of drag reduction increased significantly as the pipe diameter size was increased. The transition Reynolds number of the hydrophobic wall pipe increased slightly. At the transition, the friction factor curve increased at a lower rate than that of the smooth pipe. In accordance with the increase in Reynolds number, the experimental data for the hydrophobic wall pipe fit the experimental data for the acrylic smooth pipe. Drag reduction does not occur in the turbulent flow range. It is important to ask in considering the mechanism of drag reduction for the hydrophobic wall pipe. Why does it not occur in the turbulent flow range? The answer provides the key to clarifying the cause of apparent fluid slip. This matter is considered in more detail in Chapter 4.

The friction factors obtained using the pressure-driven-type pipeline system are presented in Fig. **3**. The dotted lines in the figure indicate the values obtained for the 6-mm-diameter pipe with hydrophobic wall in the circulation-type pipeline system. Clearly, drag reduction occurs also in the laminar flow range, and the values measured under 0, 0.1, and 0.2 MPa almost agree. Thus, it can be concluded that degassing the water has no effect on drag reduction.

The amount of drag reduction is affected significantly by the diameter of the pipe and fluid viscosity, and this phenomenon is of primary interest from viewpoint of industrial application. This relation may be inferred from Eq. (2-7) if the value of the sliding constant, β is known. However, it remains the difficulty in numerical identification of β. Since β is given the ratio of the wall shear stress and the slip velocity at the surface by Eq.(1-1) in Chapter 1, it is necessary to measure the slip velocity.

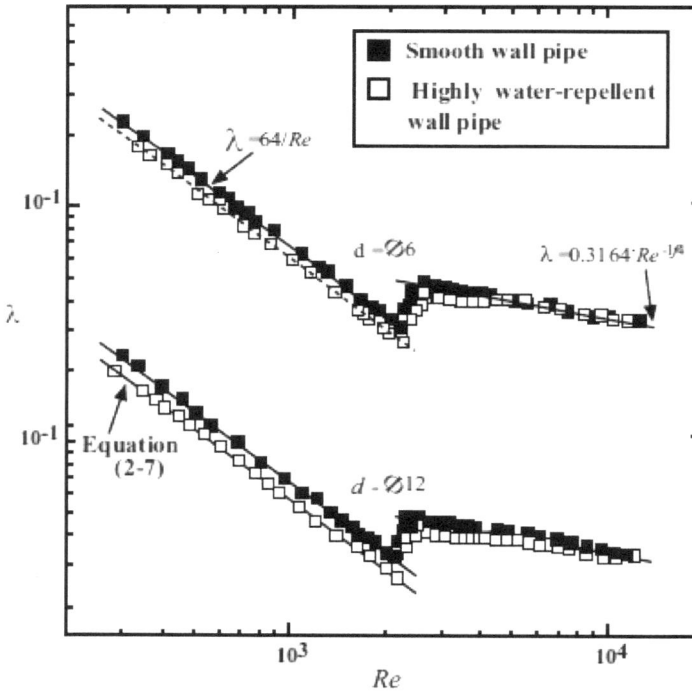

Figure 2: Friction factors.

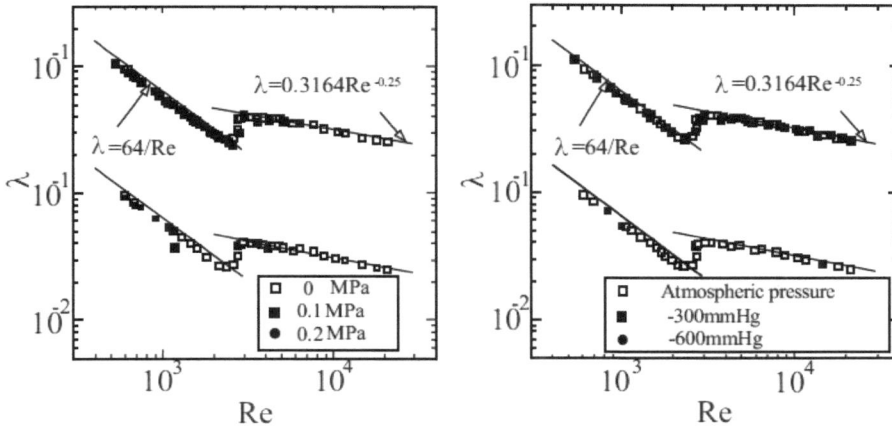

(a) Effect of adding pressure   (b) Effect of degasing of solutions

Figure 3: Friction factors in 6 mm pipe with a highly water-repellent wall.

3.2. Velocity Profile

Despite the simplicity of the flow field, few experimental data are available on the slip velocity of Newtonian fluids in a circular pipe. Fig. **4** shows the velocity profile of tap water for fully developed laminar flow through a pipe with hydrophobic wall. Experimental data were obtained using a hot-film anemometer which is a commercial product with a cone-shaped tipped type for liquid. The diameter of test pipe was 12 mm. In the figure, the velocity profile through an acrylic resin pipe is also shown for comparison under the same pressure gradient condition, $-(dp/dz)=22.6$ Pa m^{-1}. Experimental data of smooth wall pipe agree well with the solid line which is an exact solution of Navier-Stokes equation analyzing under no-slip boundary condition.

In Fig. **4**, the velocity in the hydrophobic wall pipe increases in comparison with the data of a smooth pipe. Experimental data were not obtained at the vicinity of the pipe wall because of the diameter of the support pipe of the hot film anemometer was 1.5 mm. Although the occurrence of slip velocity at the pipe wall is predicted, the line comes in Fig. **4** from extrapolation of experimental data. Needless to say, this result indicates that drag reduction occurred in this range as mentioned in Section 3-1 of this chapter.

The drag reduction ratios obtained by integrating the velocity profile in Fig. **4** and by measuring the pressure drop are 13.8% and 14%, respectively. These two values almost agree. The slip velocity was observed to increase in the two cases of tap water and glycerin solutions with an increase in the Reynolds number in the velocity profile. The wall shear stress can be calculated from the measured pressure loss because they equal out in fully developed flow in a pipe. As described above, we can experimentally determine the physical constant β from the graph which shows the relationship between the wall shear stress τ_w and the slip velocity u_s.

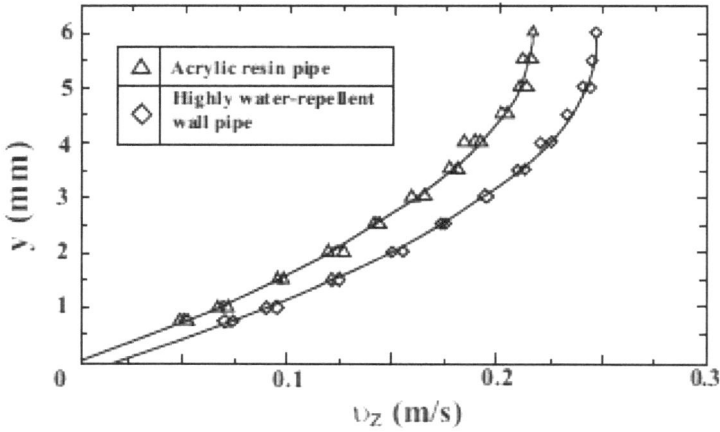

Figure 4: Velocity profiles of tap water in pipe with a highly water-repellent wall.

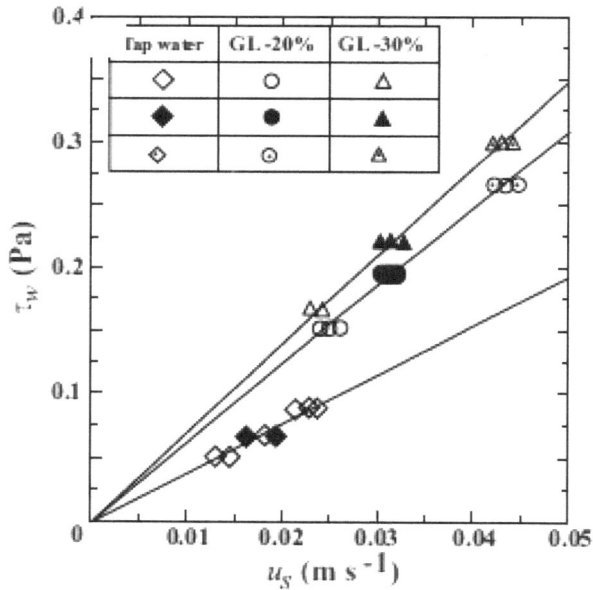

Figure 5: Relationship between wall shear stress and slip velocity.

The relationship between τ_w calculated using experimental data on the pressure, and u_s obtained by extrapolating from the velocity profile, is presented in Fig. **5**. The figure shows that the slip velocity is directly proportional to the wall shear stress; the sliding constant β is a constant value since the gradient gives the value. These results are very important, because for the first time they experimentally confirm Eq. (1) in Chapter 1 based on the Navier's hypothesis. It was clarified that the sliding constant β increases slightly with an increase of in viscosity, and has the values 3-7 Pa s/m in this case. We might expect that the friction factor could be calculated similar to Eq. (2-7) from estimating the value of the sliding constant for the flow condition. In Fig. **2**, the results of Eq. (2-7), calculated by substituting the value of β, are represented by solid lines. The analytical result agrees well with the experimental results in laminar flow range.

Since the product of Reynolds number and the friction factor becomes the function only for the non-dimensional parameter $S = (\mu/a\beta)$, the comparison between the analytical result and the experimental data is also expressed in a relatively simple curve. Fig. **6** shows the result. Good agreement is provided between them as is obvious. However, it is not sufficient for the comparison with other flow system size because the value of β is obtained only for the case of 6 mm diameter pipe. The value of β depends on not only the properties of hydrophobic surface and the physical constants of liquid, but also the flow system size, *e.g.* the pipe diameter.

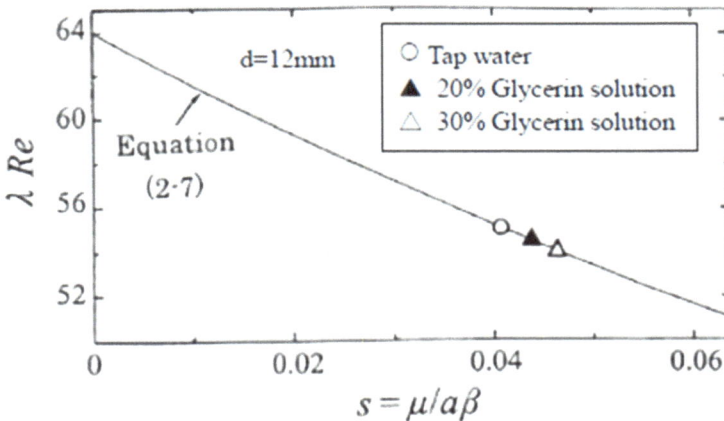

Figure 6: Friction factors for the slip flow.

The coated wall used for the experiment has 130° to 140° of a contact angle with water, and it is larger than that of a regular smooth PTFE surface. According to the result that the water does not slip at regular smooth PTFE surface, the contact angle can be one of the criteria for estimating the slip velocity. But, the experimental data about the slip velocity are lacking. One problem is obtaining meaningful measurements of the sliding constant value by laboratory instrument. This discussion indicates that the phenomena underlying the observed behavior of Newtonian fluid slip in a pipe flow are still far from being fully understood from a macroscopic viewpoint. On the other hand, the mechanism of the fluid slip at the highly water-repellent wall is more complex. As is evident in Fig. **4a** in Chapter 1, the surface has many fine grooves and fractal structure. There is a great chance that the interface of air phase exists in a groove. Thus, it will be necessary to use other analytical method to clarify the flow behavior around a blunt body namely a sphere or cylinder *etc.*, even if we can calculate the friction factor using the Navier's hypothesis.

SUMMARY

Experiments showed that laminar drag is reduced for Newtonian fluids that flow in a pipe with a highly water-repellent wall, as demonstrated by measurements of the pressure loss and of the velocity profiles. The drag reduction ratio is at most approximately 20% for tap water, depending on the flow-system size and viscosity. The reason for this phenomenon is the apparent fluid slip that occurs at a highly water-repellent pipe wall. The predominant cause of this slip is the presence of an air-liquid interface at the wall. In a large flow system, the extent of this interface must be maintained to obtain fluid slip. Additionally, surface tension prevents tap water or glycerin solution from making contact with the surface of fine grooves. The air between the liquid surface and the groove in the wall plays an important role in producing fluid slip. This can be accounted for by the fact that drag reduction does not occur in turbulent flow or with surfactant solutions that have a lower surface tension than tap water.

An analytical result for the friction factor was obtained using Navier's hypothetical equation, where a value for the slip velocity at a pipe wall is assumed. The result agrees well with measurements, with the sliding coefficient ranging from $\beta = 1$ to 10 (Pa· s/m).

REFERENCES

[1]   Goldstein, S., (1965), Modern Developments in Fluid Dynamics, Vol.2, 676-6890, Dover
[2]   Watanabe, K., (2002), Fluid Dynamics-Flow and Fluid Drag- (in Japanese), 49, Marzen
[3]   Langhaar, H. L., (1942), "Steady flow in the transition length of a straight tube", J. Appl. Mech. 9, Trans. ASME, **64**, A55-58,
[4]   Watanabe, K., Yanuar and Udagawa, H., (1999), "Drag reduction of Newtonian fluid in a circular pipe with a highly water-repellent wall", J. Fluid Mech., **381**, 225-238.

Laminar Drag Reduction, 2015, 21-27

CHAPTER 3

Non-Newtonian Fluids Flow in a Circular Pipe

Abstract: High molecular weight polymer solutions or surfactant solutions are typical examples of complex fluids. They exhibit nonlinearity in viscosity and drag reduction occurs in the turbulent pipe flow even if the concentration is very dilute. They are generally known as non-Newtonian fluids. By applying a hydrophobic wall pipe to reduce drag on the flow of polymer solutions, a flow system was constructed, wherein drag reduction was obtained in both the laminar and turbulent flow ranges. In discussions of Newtonian fluid in Chapter 2, we dealt with apparent slip flow analytically using Navier's hypothesis, and the result was compared with the experimental findings of the friction factor. The experimental results of PEO (polyethylene oxide) aqueous solutions with a concentration range of 30-1000 ppm and the analytical result for the friction factor of a power-law fluid with fluid slip were analyzed by applying the modified boundary condition on fluid slip, as described in this chapter.

Keywords: Drag reduction, laminar flow, complex fluid, turbulent drag reduction, pipe flow, friction factor, highly water-repellent wall, slip velocity, power-law model, polyethylene-oxide solutions, surfactant solutions, contact angle.

1. INTRODUCTION

Non-Newtonian fluids flow in a pipe appears frequently in the flows of fiber formation, plastic molding processes and food processing. In the analysis of the non-Newtonian fluids flow, it is necessary to determine a rheological equation of the fluids for the flow behavior. The stress of the non-Newtonian fluids is expressed as a function of velocity gradient, although it is proportional to the velocity gradient in Newtonian fluids by Newton's law of viscosity. If the rheological equation of the fluid is determined or assumed, the velocity profile and the drop in pressure of steady laminar flow in a pipe can be derived by simple theoretical analysis paralleling the approach used for obtaining the relationships for Newtonian fluids. Although some rheological equations have been previously proposed for the rheological behavior, the most simple and convenient equation for this analytical approach is that for a power-law fluid, in which two physical constants must be determined experimentally.

We start our discussion with the experimental results [1] for the friction factor of a high molecular weight polymer solution in a pipe, which was presented using a power-law fluid. We deal next with the analysis assuming the slip flow at the pipe

Keizo Watanabe

wall. Finally, a comparison between the experimental and analytical results is presented.

2. EXPERIMENTAL CORRELATIONS

The experimentally determined friction factors of PEO 15 (polyethylene oxide) dilute and highly concentrated solutions are shown in Figs. **1a** and **b**, respectively. Here, the solid lines indicate $\lambda = 64/\text{Re}$ for laminar flow and Blasius equation for turbulent flow. Virk's the maximum drag reduction friction factor curve [2] are also shown. The polymer solutions have a significant non-Newtonian viscosity in concentration over 200 ppm. Thus, a power-law model is used to organize the experimental data. The rheological equation of a power-law model is given for a steady laminar pipe flow on a cylindrical coordinate as follows,

$$
\tau_{rz} = K \left(-\frac{d\upsilon_z}{dr} \right)^n , \tag{3-1}
$$

where, K and n are the consistency constant and power, respectively. These are the physical constants experimentally determined *via* the flow curve. υ_z is the velocity in the axial direction. In Fig. **1b**, Re^* is the power-law Reynolds number given as $\text{Re}^* = \left(8n^n \rho d^n u^{-2-n} / 2^n (3n+1)^n K \right)$.

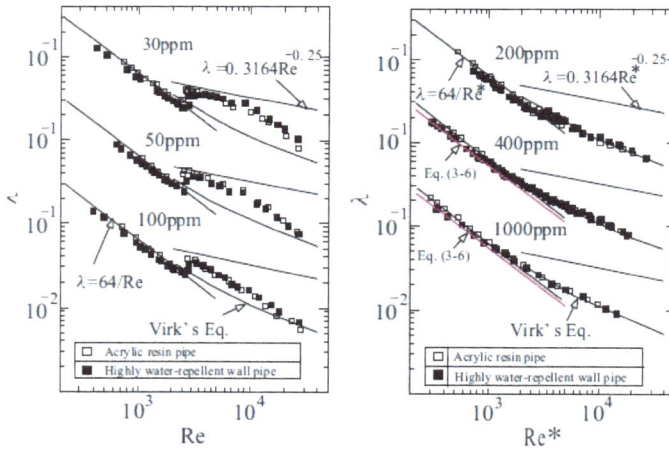

(a) Dilute solutions   (b) Highly concentrated solutions

Figure 1: Friction factor of PEO solutions.

Experimental data are obtained using the pressure-driven pipeline system consisting of a pressure tank and a compressor. The pipe tested, having a hydrophobic wall of the same type described in Chapter 2, is 6 mm in diameter and 430 mm in test-section length. Polymer solutions passed once through the pipeline of the experimental apparatus are discarded to prevent the degradation of solutions.

Fig. **1** shows clearly that drag reduction in laminar flow occurs with polymer solutions in the hydrophobic wall pipe, but not in the smooth acrylic resin pipe. The drag reduction ratio is approximately 11%-15% and increases slightly with an increase in concentration. However, the difference vanishes in the turbulent flow range, and the friction factor fits Virk's equation with an increase in the Reynolds number. The variation is clarified by rearranging the experimental data as shown in Fig. **2**. As mentioned in Chapter 2, for Newtonian fluids, the friction factor λ of the hydrophobic wall pipe decreases comparable to that of the smooth acrylic resin pipe. However, this trend is not observed for polymer solutions of any concentration in the measured Reynolds number range. The tendency was particularly noted with highly concentrated solutions of 400 ppm and 1000 ppm. The probable reason is that the flow behavior on non-Newtonian fluids is unstable at the transition range from laminar to turbulent compared with that of Newtonian fluids. In particular, this instability in the laminar flow range occurs in experimental data measured using apparatus with an open outlet pipe-line, as shown experimentally by Hasegawa & Tomita [3] for PEO solutions. It is likely that the fluctuation of pressure at the wall cannot be neglected for slip velocity, thereby resulting in this unstable.

Surfactants are drag reducing additives, as listed in Table **1**, and the solutions do not induce degradation in a circular pipe-line system with a centrifugal pump. Thus, they are practically applied for district heating and cooling pipeline systems in order to save energy. On the other hand, it is important to clarify the friction factor experimentally in order to consider the mechanism of fluid slip at the hydrophobic wall because the surface tension is very low compared to that of tap water. Friction factors of surfactant solutions (Ethoquad O/12) are shown in Fig. **3**. Surfactant solutions exhibit the flow behavior of an SIS (shear-induced structure), the apparent viscosity increases rapidly with a decrease in shear rate. For this reason, it is at present difficult to obtain the rheological equation. Therefore, the Reynolds number is calculated using the viscosity of tap water in the figure.

Drag reduction of surfactant solutions observed in turbulent flow range, is not the laminar drag reduction with fluid slip. Experimental data of surfactant solutions in highly water repellent wall pipes almost fit to that of smooth wall pipe as shown in Fig. **3**. Fluid slip does not occur in laminar flow of surfactant solutions. Thus, the laminar drag reduction is influenced by a surface tension of liquids. However, an interesting result from Fig. **3** is that the difference in the friction factor of surfactant solutions of the smooth acrylic resin pipe and the hydrophobic wall pipe is not recognized in both the laminar and turbulent flow ranges. Laminar drag reduction does not occur. In other words, the apparent slip velocity does not occur at the hydrophobic wall for surfactant solutions although it does occur in the case of high molecular polymer solutions. This result is compelling evidence in support of the occurrence of apparent slip velocity in that if an air-liquid interface is present at the wall surface, slip velocity is observed. This closely agrees with observation of the microphotographs in Figs. **3, 4a, b & c** in which the contact angle of surfactant solutions is less than that of tap water or polymer solutions.

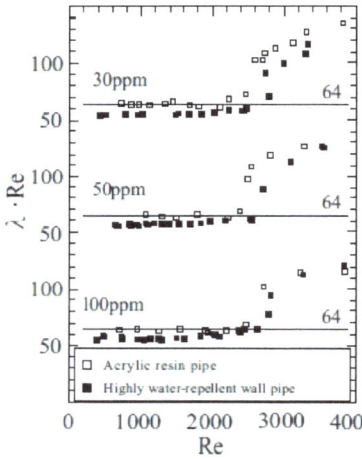

Figure 2: ($\lambda \cdot$Re) *versus* Re.

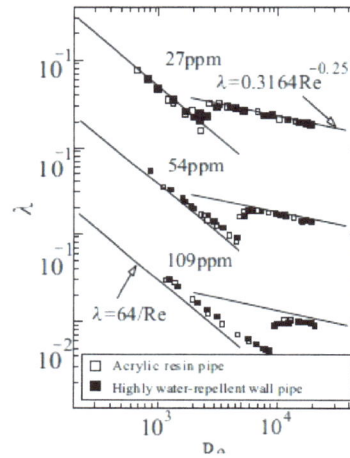

Figure 3: Friction factor of surfactant solutions.

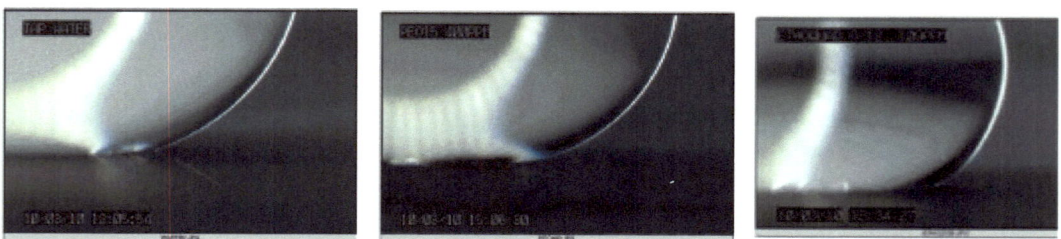

(a) Tap water	(b) PEO15 400 ppm solution	(c) Surfactant solution

Figure 4: Shape of a droplet on hydrophobic wall.

It may be that the decreasing surface tension of the liquid degrades the functionality of the hydrophobic wall such that the contact angle of the surface decreases and the water repellency varnishes. The behavior of the interface between air and liquid at the hydrophobic wall with fine grooves will be discussed in Chapter 4 with the visual results of flow in a duct.

3. ANALYTICAL APPROACH

The stress of the non-Newtonian fluids is expressed as a function of the shear rate, although it changes in proportion to the shear rate in Newtonian fluids as mentioned above. The nonlinearity of viscosity is represented by an approximate deconvolution model as known as the rheological equation for power-law fluids.

We start the discussion by analytically considering fluid slip, which is important in the interpretation of power-law fluids in pipe flow. The fundamental momentum equation for steady fully developed laminar flow in a pipe is given as follows;

$$\frac{1}{r}\frac{d}{dr}\left(r\tau_{rz}\right) = -\left(\frac{dp}{dz}\right) \tag{3-2}$$

where τ_{rz} and p denote the shear stress and pressure, respectively. r and z denote the cylindrical coordinate and z is the flow direction.

The power-law fluid rheological model given in Eq. (3-1) is substituted in Eq. (3-2). By integrating Eq. (3-2) and substituting fluid slip boundary condition at the pipe wall, we obtain the following equation for the velocity profile

$$\upsilon_z = \left(-\frac{dp}{dz}\right)^{\frac{1}{n}} \frac{na^{\frac{n+1}{n}}}{2^{\frac{1}{n}}K^{\frac{1}{n}}(n+1)}\left\{1-\left(\frac{r}{a}\right)^{\frac{n+1}{n}}\right\} + u_s \tag{3-3}$$

where, a and u_s are the radius of the pipe and the slip velocity, respectively. Although the effect of the slip velocity is assumed to be linear to the wall shear stress in the case of Newtonian fluids, this assumption may not apply in the case of power-law fluids. Using Navier's hypothesis, we assume the slip velocity u_s in Eq. (3-3) to be as follows

$$\tau_w = \beta u_s^{\ m} \tag{3-4}$$

where, m is a physical constant and is determined empirically.

Thus, the velocity profile of a power-law fluid is calculated using Eq. (3-4)

$$\upsilon_z = \left(\frac{\Delta p}{2KL}\right)^{\frac{1}{n}} \frac{nR^{\frac{n+1}{n}}}{n+1}\left\{1-\left(\frac{r}{R}\right)^{\frac{n+1}{n}}\right\} + \left(\frac{R\Delta p}{2\beta L}\right)^{\frac{1}{m}} \tag{3-5}$$

where $(\Delta p / L) = (-dp / dz)$ and Δp and L are the pressure drop and the length of pipe, respectively.

We also assume $m = n$ for the parameter m in Eq. (3-4) by considering the dependence of viscosity for a sliding constant.

Finally, the friction factor of power-law fluids with fluid slip is obtained as follows

$$\lambda = \frac{64}{\mathrm{Re}^*}\frac{1}{\left[1+\dfrac{(3n+1)K^{\frac{1}{n}}}{nR\beta^{\frac{1}{n}}}\right]^n} \tag{3-6}$$

where, Re^* is the power-law Reynolds number as given by Eq. (3-1). Equation (3-6) agrees with Eq. (2-7) in the case of Newtonian fluids, such as $n = 1$.

In order to calculate the friction factor using Eq. (3-6), it is necessary to determine the value of the sliding constant β. We can estimate the value from the experimental result of the friction factor shown in Fig. **1b**. In the figure, under the line of $\lambda = 64 / \mathrm{Re}^*$ solid lines calculated by substituting the value of β are indicated. The values calculated back from the figure are $\beta = 9.7$ Pa·s/m and 12.5 Pa·s/m for 400 and 1000 ppm PEO 15 solutions, respectively. The results agree with the experimental results at Reynolds number less than 800 for a PEO15 solution of 400 ppm. With increases in the concentration, the experimental results deviate from the analytical result in the low Reynolds number range. Thus, the relationship between the wall shear stress and the slip velocity for PEO solutions can be approximated to the power law in this range.

SUMMARY

Non-Newtonian fluids flow behavior applies to problems in the processing of food, crude oil and polymer products. It is well known that the rheological behavior of such fluids does not follow the Newton postulate that the applied stress is strictly proportional to the velocity gradient for a simple shear flow. Although there are many non-Newtonian fluids, the flow behaviors of polymer and surfactant solutions has recently been considered from the view point of drag reduction in turbulent flow in the pipeline systems. In this chapter, fluid-slip phenomena in such solutions were examined experimentally in pipeline systems. An interesting conclusion was that laminar drag reduction does not occur for surfactant solutions of a circular pipe that has a highly water-repellent wall, although it does occurs with polymer solutions. The difference between surfactant and polymer solutions with regard to the surface tension at the gas—liquid interface at the highly water-repellent wall explains the apparent fluid slip. Thus, the experimental results described in this chapter provide insight for explaining laminar drag reduction and fluid slip. An analytical result for the friction factor of high concentration polymer solutions was obtained using a power-law fluid model and a deformation model for Navier's hypothetical equation. The result agrees well with measurements qualitatively.

REFERENCES

[1]   Watanabe, K and Udagawa, H, (2001), "Drag Reduction of Non-Newtonian Fluids in a Circular Pipe With a Highly", AIChE J., **26** pp. 729-739.

[2]   Virk, P. S., (1975), "Drag Reduction Fundamentals", AICHE Journal, **21** pp. 625-656.

[3]   Hasegawa, T., and Tomita Y., (1974), "A Study on Anomalous Turbulent Flows of Non-Newtonian Fluids" (3rd Report, Experimental and Analysis in Transitional Region), Bull. of JSME, **17** 103, pp.73-82.

Chapter 4: *Flow in a Duct*

Flow in a Duct

Abstract The laminar flow in a duct can be found by the exact solution of the Navier-Stokes equation. An experimental apparatus that changes the flow using a detachable channel side plate can be used to measure the velocity profile or pressure drop in order to find the dominant determiner of the wall's characteristics. In this section, the mechanism of the apparent fluid slip is discussed by comparing the experimental test results from two ducts having hydrophobic walls with different structural characteristic but almost the same water-repellency characteristics. Different kinds of hydrophobic walls were produced for duct wall, and the laminar drag reduction was examined by measuring the pressure drop.

Keywords: Drag reduction, laminar flow, duct flow, exact solution, friction factor, velocity profile, highly water-repellent wall, slip velocity, fractal surface, prototype surface, flow visualization, gas-liquid interface.

1. INTRODUCTION

In Chapter 2 showed that the apparent fluid slip of a Newtonian fluid occurs at a hydrophobic wall with many fine grooves, of which the contact angle is about 140 degrees, although it is well known that the no-slip boundary condition is satisfied at a normal solid surface for a Newtonian fluid. Because a rectangular or a square duct consists of four flat side plates, we can make a duct with a removable side plate whose surface has various shapes. Thus, it is convenient to examine the drag reduction phenomenon experimentally by using a pipeline system with these ducts. Some experimental results for the friction factor of a duct are obtained and discussed by a comparison with analytical results. By using examples from these experimental results, some drag reducing walls for a liquid are fabricated and the friction factor is obtained. To determine the mechanism of the apparent fluid slip at a hydrophobic wall surface, it will be necessary to clarify the flow behavior in proximity to the wall using flow visualization. In this chapter, the flow behavior of tap water near the wall proximity is shown by means of a flow visualization system with a microscope.

2. EXPERIMENTAL APPROACH

Experiments [1, 2] were carried out to measure the pressure drops of a rectangular duct 2a × 2b = 20mm×10mm, and a square duct (15mm×15mm). These had aspect ratio of 0.5 and 1.0, respectively, where $\varepsilon = (2b/2a)$. The velocity profile for the square duct was measured using a hot film anemometer. The friction factors for the tap water are shown in Fig. **1**. In Fig. **1**, the solid line is $\lambda_c = 64 / \mathrm{Re}$ for a circular pipe flow, and the dotted lines are the analytical results [3] for Newtonian fluids in a duct under the no-slip boundary condition. Values of $\lambda = 0.9716\lambda_c$ and $\lambda = 0.887\lambda_c$ were found for ducts with $\varepsilon = 0.5$ and $\varepsilon = 1.0$, respectively. The experimental data for smooth wall ducts made of acrylic resin agreed well with the analytical results in the laminar flow region. The experimental data shifted to the turbulent flow range at Re=2.0-2.2×10³ as was pointed out in the experimental results of Hartnett *et al.* [4]. Experimental data for a hydrophobic wall duct decreased in parallel with the data for a smooth wall duct and the transition to turbulent flow was delayed. In turbulent flow range, these data fit the data for a smooth wall duct. However, these data fit the data for a smooth wall duct. This phenomenon was similar to that of a circular pipe with hydrophobic walls including the laminar drag reduction. Drag reduction ratios of about 15% and 22% were found for the rectangular and square ducts, respectively.

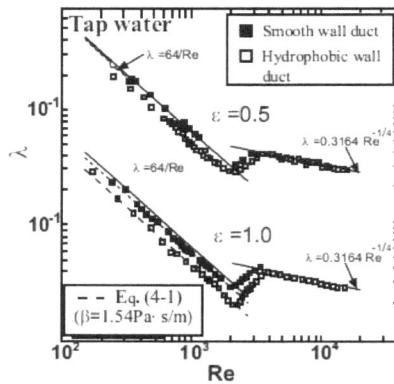

Figure 1: Friction factor of ducts.

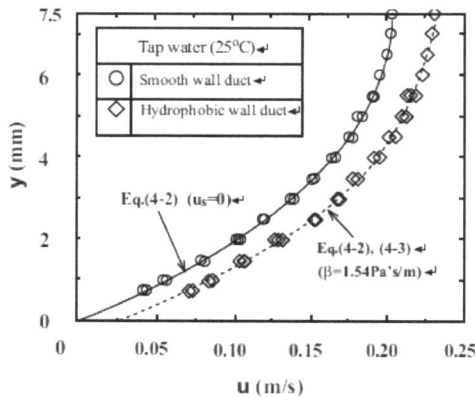

Figure 2: Velocity profiles of tap water in a square duct.

Fig. **2** shows the comparison between the velocity profiles of the flow in a square duct with smooth walls and hydrophobic walls. In Fig. **2**, the pressure gradient is $(-dp/dz)=10.8$ Pa/m in both cases. Fluid slip can be considered occurs at a hydrophobic wall duct, although the no-slip flow condition is satisfied in the smooth wall duct made of acrylic resin. Because these experimental data of velocity profiles in Fig. **2** were obtained under same pressure gradient as described above, we can calculate the drag redaction ratio by the graphical integration of Fig. **2**. It was about 23.4% in the drag reduction ratio. This value, as a matter of course, almost agrees with the drag reduction ratio obtained from Fig. **1**. On the other hand, two square ducts with different kinds of hydrophobic walls were tested against the pressure loss and velocity profile of tap water in order to clarify the effect of the surface structure on the laminar drag reduction. Fig. **3a** and **3b** show micrographs of the wall. Fig. **4a** and **b** show the shapes of water drops on these walls. Although both walls A and B, were made of PTFE

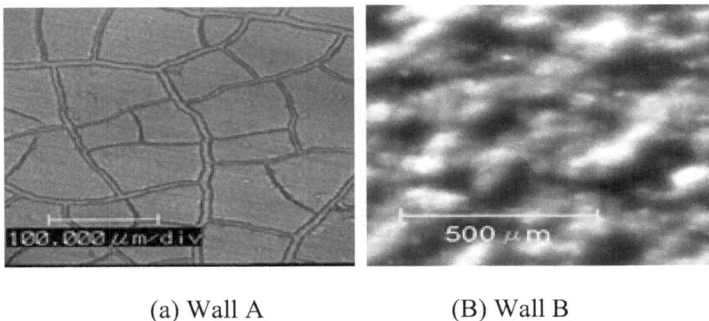

(a) Wall A   (B) Wall B

Figure 3: Surface of hydrophobic walls. Note that many fine grooves are formed on Wall A.

(polytetrafluoroethylene) and behaved as highly water-repellent walls because of the contact angle, which was about 150°, there were many fine grooves at the surface of wall A, as shown in Fig. **4a**, whereas wall B had no fine grooves even though it was a fractal, as shown Fig. **4b**. As shown by the droplet shape in Fig. **4**, both of these hydrophobic walls had a highly water repellent property. Nevertheless, they had different surface structures. Experiments were carried out to measure the pressure loss and velocity profile of tap water in the square duct (15mm × 15mm). The velocity profile was measured by a PIV technique by applying a hydrogen bubble method. A 50 μm diameter platinum wire was used for the negative electrode in this method. The flow behavior of the test fluid near the surface of wall A was observed with 3000 ×magnification using a microscope set under the test section. A video recorder with a light spot recorded the flow patterns. Then, flow visualization results of the tap water and an aqueous solution of oreyl-bishydroxyethl-methy-ammonium $(C_{18}H_{35}N(C_2H_4OH)_2CH_3Cl$, trade name: Ethoquad O/12) at a concentration 200ppm were obtained to clarify the mechanism for the laminar drag reduction phenomenon. Sodium salicylate (NaSal) was added to the solution as a counterion. The concentration ratio of sodium salicylate to Ethoquad O/12 was set at a molar ratio of 1:1. It is well known that the drag reduction occurs in surfactant solutions at the turbulent flow region because of turbulence modification.

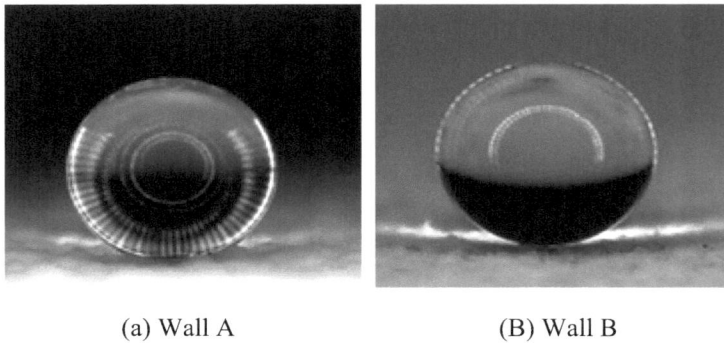

(a) Wall A   (B) Wall B

Figure 4: Shape of a droplet at the hydrophobic wall.

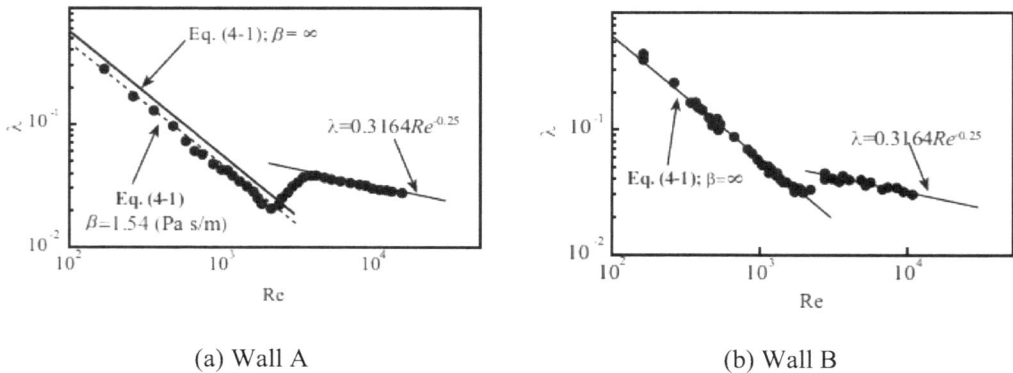

(a) Wall A   (b) Wall B

Figure 5: Friction factor of tap water in duct.

Experimental results for the friction factor of the square ducts with walls A and B are shown in Fig. **5a** and **b**, respectively. The solid lines denote the exact solution [1] and Blasius' equation for laminar and turbulent flow ranges obtained using the no slip boundary condition, respectively. In the duct with wall A, the reduction in the friction factor occurs in the laminar flow range although the experimental data agreed well with Blasius' formula in a turbulent flow range as shown in Fig. **5a**. It has drag reduction ratio of approximately 22%. However the drag reduction phenomenon does not occur for the duct with wall B as shown in Fig. **5b**. The phenomenon is verified by the experimental data of the velocity profile as shown in Fig. **6**. A comparison of the data shows that the slip velocity occurs at wall A but does not occur at wall B.

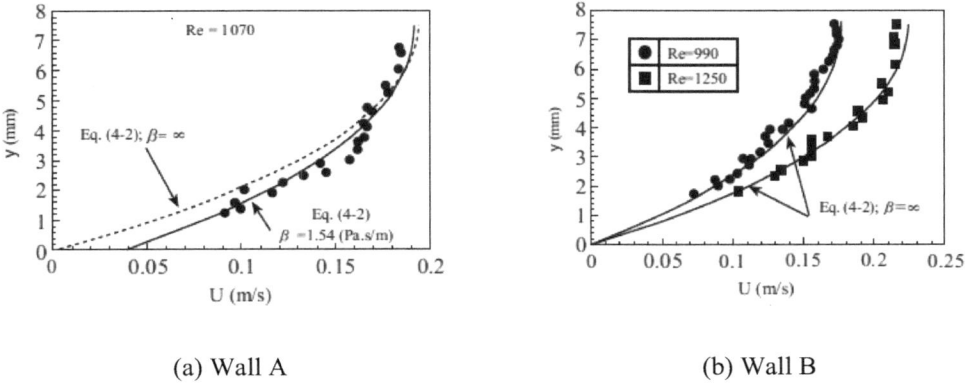

(a) Wall A   (b) Wall B

Figure 6: Velocity profile of tap water in duct.

In general, it is meaningful to draw a comparison between these experimental and analytical results in order to grasp the phenomenon. By using Eq.(1-1) for fluid slip boundary conditions, the friction factor of a duct is derived as Eq.(4-1) from the Navier–Stokes equation.

$$\lambda = \frac{64}{Re} \frac{3/2}{(1+\varepsilon)^2 \left[\left(1 + \frac{3\mu}{\beta b}\right) - \frac{192}{\pi^5} \varepsilon \sum_{n=1,3,5,\cdots}^{\infty} \left\{ \frac{1}{n^5} \left(1 + \frac{n^2\pi^2\mu}{4\beta b}\right) \tanh\left(\frac{n\pi}{2\varepsilon}\right) \right\} \right]} \qquad (4\text{-}1)$$

where β is the sliding coefficient, ε is the aspect ratio of the duct, and $2a$ and $2..$are the width and height of the duct, respectively. With the limit for Eq. (4-1), $\beta \rightarrow \infty$ we obtain the exact solution [1] of the friction factor for a duct with no fluid slip wall. Re is the Reynolds number defined as $(4\rho m U_m / \mu)$, and $\rho,, \mu$ and U_m are the fluid density, hydraulic mean depth, viscosity, and mean velocity, respectively. Examples of the calculation results are shown in Fig. **7** for the square and rectangular ducts. If the value of the sliding constant, β, is extrapolated from the experimental results in Fig. **1**, it is found to be β=3.04 and 1.54 Pa· s/m for ε =0.5 and ε =1.0, respectively as shown by the white and black circles in Fig. **7**. The results are also shown in Figs. **1** and **5**. In these figures, the dotted line denotes the result of Eq. (4-1), where the value of β=1.54 Pa· s/m is inserted.

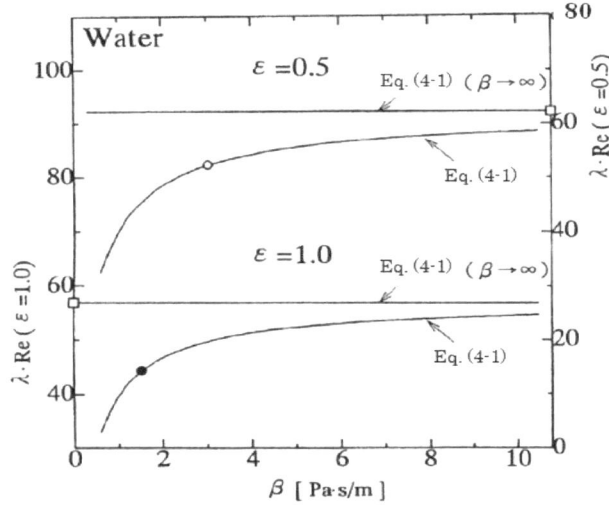

Figure 7: Calculation results of Eq. (4-1).

The experimental data for the friction factor agree well with the results from the extrapolation. As described in Chapter 2, it is 1~10 of the order of β in the case of the hydrophobic wall pipe, and β=1.54 Pa· s/m is about same order with it. On the other hand, the velocity profile in the duct is presented in Fig. **6**. The experimental results indicate that no reduction in drag is likely to occur in the absence of a gas-liquid interface. The analytical results of the velocity profile were obtained during the course of the friction factor analysis.

The axial velocity profile is given as follows:

$$u = \frac{1}{\mu}\left(-\frac{dp}{dz}\right)\left[\frac{y}{2}(2b-y) - \frac{16b^2}{\pi^3}\sum_{n=1,3,5,\cdots}^{\infty}\frac{1}{n^3}\sin\left(\frac{n\pi y}{2b}\right)\frac{\cosh\left(\dfrac{n\pi x}{2b}\right)}{\cosh\left(\dfrac{n\pi a}{2b}\right)}\right] + u_s \quad (4\text{-}2)$$

where, y is the distance from the wall of the duct, and u_s is the slip velocity given by

$$u_s = \frac{b}{\beta}\left(-\frac{dp}{dz}\right)\left[1 - \frac{8}{\pi^2}\sum_{n=1,3,5,\cdots}^{\infty}\frac{1}{n^2}\frac{\cosh\left(\dfrac{n\pi x}{2b}\right)}{\cosh\left(\dfrac{n\pi a}{2b}\right)}\right] \quad (4\text{-}3)$$

As shown in Fig. **6**, the experimental data of the duct with wall A agree well with the solid line. This fact indicates as mentioned above, that the slip velocity occurs at a highly water-repellent wall, which also causes a drag reduction to occur. For the duct with wall B, the experimental data were expected to fit the solid lines obtained under the no-slip boundary condition as shown in Fig. **6b**.

Consequently, the experimental results for the friction factor and velocity profile of flow in two ducts with different wall surface structures clarified that the drag reduction phenomena based on the slip velocity do not occur at a surface with no fine grooves even if the surface is a hydrophobic wall that is highly repellent. Thus, the grooves fulfils the most important role for the occurrence of the apparent slip velocity at the wall. The experimental results that the drag reduction does not occur if the gas–liquid interface is not held at the surface and the apparent slip velocity does not occur, were confirmed by other experiment results that the friction factor of a surfactant solutions flowing through a pipe with wall A fit that of tap water in an acrylic resin pipe. In other words, it does not function as a highly water-repellent wall for the flow of surfactant solutions. Because the value of the surface tension of surfactant solutions is approximately half the value of tap water, it can be considered that these fine grooves were filled with the solutions; that is, the surface has a degassed condition. Thus, the apparent slip velocity does not occur at wall B even if it has a highly water repellent surface.

3. PROTOTYPE DRAG REDUCING WALL

By referring to these experimental results, we can make a drag reducing wall. The necessary and sufficient conditions the wall are that it has to have a fractal surface and be highly water repellent. Fig. **8a, b & c** show the typical roughness pattern of the surface [5] and the etching applied to a silicon substrate. The groove depth and area are d=20 μm and 15%, respectively. The width values are w = 10 and 5 µm, and the pitch values are p=33 and 66 µm in the pitch. Dimension of random groove of Fig. **8c**, is shown in detail Fig. **3** in Chapter 1.

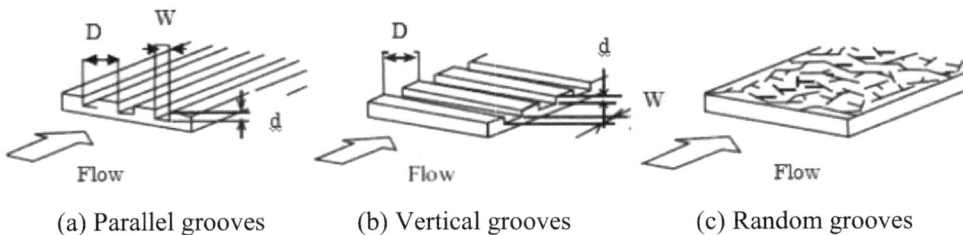

(a) Parallel grooves   (b) Vertical grooves   (c) Random grooves

Figure 8: Roughness dimension of surface.

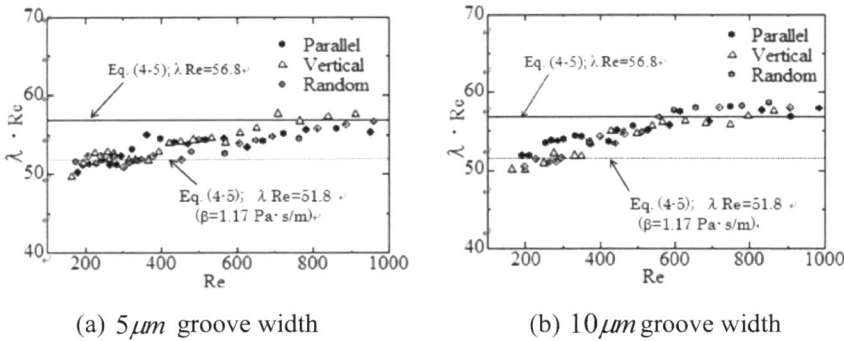

(a) $5\mu m$ groove width   (b) $10\mu m$ groove width

Figure 9: Friction factor of a rectangular duct with artificial grooves.

A vertical groove where the direction is perpendicular to the flow is also made. A wall was coated with by PTFE to fabricate a drag reducing hydrophobic test wall. Experiments were carried out to measure the pressure loss of a square duct with a hydrophobic wall on one side in a horizontal pressure-driven pipeline system with a constant over flow head tank. The length of the test section for the pressure measurement was 250 mm, and the test fluid was tap water. The experimental results for the friction factor are shown in Fig. **9a** and **b**. Drag reduction was observed in both cases at the low Reynolds number range. However, the experimental data increases with an increase in the Reynolds number, after which the effect at Re$\geqq$600 for the 10 µm-groove-width wall. The 5 µm- groove-width wall covered a large range width wall. Trends of this type may be anticipated with the flow visualization results for the gas-liquid interface behavior. It is interesting to compare the results of Fig. **9a** and **b** with the analytical results for the flow in a duct with one side slip wall. In Fig. **10**, fluid slip occurs only at $y = 2b$, and the no slip boundary condition is satisfied for the other boundary. The velocity profile is obtained by analyzing to the Navier-Stokes equation. The following equation is then obtained by inserting this profile into Eq.(2-4). μ and β in the equation are viscosity and the sliding constant, respectively.

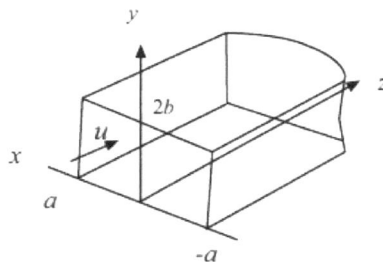

Figure 10: Coordinate system for a duct.

$$u = \left(-\frac{dp}{dz}\right)\frac{16a^2}{\mu\pi^3}\sum_{n=1,3,5,\cdots}^{\infty}\frac{(-1)^{\frac{n-1}{2}}}{n^3}\cos\frac{n\pi x}{2a}\left[-\left(1-\cosh\frac{n\pi y}{2a}\right)+\left(1-\cosh\frac{n\pi b}{a}\right)\frac{\sinh\frac{n\pi y}{2a}}{\sinh\frac{n\pi b}{a}}\right.$$

$$\left.+\frac{1-\cosh\frac{n\pi b}{a}}{\frac{2a}{n\pi}\frac{\beta}{\mu}\sinh\frac{n\pi b}{a}+\cosh\frac{n\pi b}{a}}\frac{\sinh\frac{n\pi y}{2a}}{\sinh\frac{n\pi b}{a}}\right]$$

$$(4\text{-}4)$$

The calculation results are shown in Fig. **11**. In the no-slip flow case, the velocity profile is calculated by inserting $\beta=10^3$(Pa· s/m). For the slip flow case, the same value of $\beta=1.54$ Pa· s/m used in Fig. **6a** is used for the calculation under the same pressure gradient. It becomes a slightly bellied profile compare to that of the velocity with no fluid slip. From Eq. (4-4), the friction factor is obtained and the result is given as the flow equation:

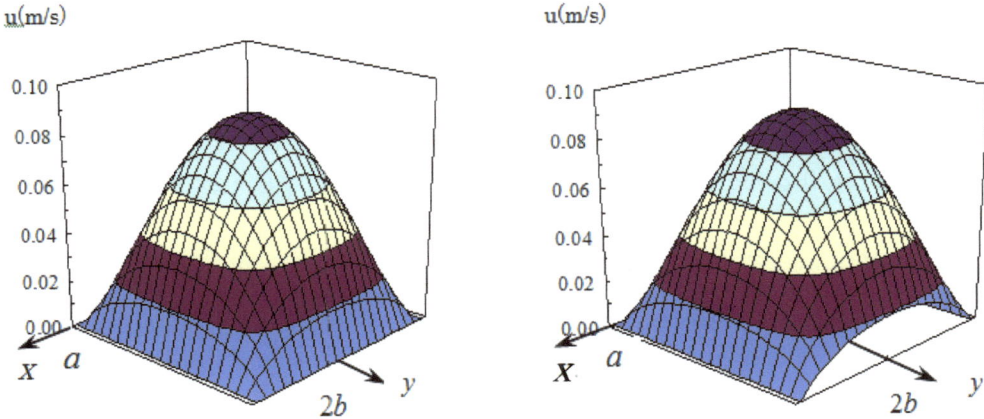

$$a=b=6mm \quad dp/dz=8.27(Pa/m)$$

(a) No-slip flow   (b) Slip flow at one side wall

(Re=500, $\beta=10^3$ Pa· s/m)   (Re=546. $\beta=1.54$ Pa· s/m)

Figure 11: Computational results of velocity profiles in rectangular duct.

$$\lambda = \frac{64}{\mathrm{Re}} \frac{\pi^4 \varepsilon^3}{64(1+\varepsilon)^2} \frac{1}{\sum_{n=1,3,5,\cdots}^{\infty} \frac{1}{n^4}\left[-\varepsilon + \frac{1}{n\pi}\left\{2\frac{\cosh n\pi\varepsilon - 1}{\sinh n\pi\varepsilon} - \frac{\frac{(\cosh n\pi\varepsilon - 1)^2}{\sinh n\pi\varepsilon}}{\frac{2a}{n\pi}\frac{\beta}{\mu}\sinh n\pi\varepsilon + \cosh n\pi\varepsilon}\right\}\right]}$$

$$(4\text{-}5)$$

A comparison of this result with the sliding constant β =1.17 Pa· s/m is shown in Fig. **9**. This is shown as the dotted lines in Fig. **9**. Because the prediction for the value of β is given by the experimental data for the friction factor, it is commonly assumed that the gas–liquid interface is held at the wall when liquid flows.

→ Flow   → Flow   → Flow

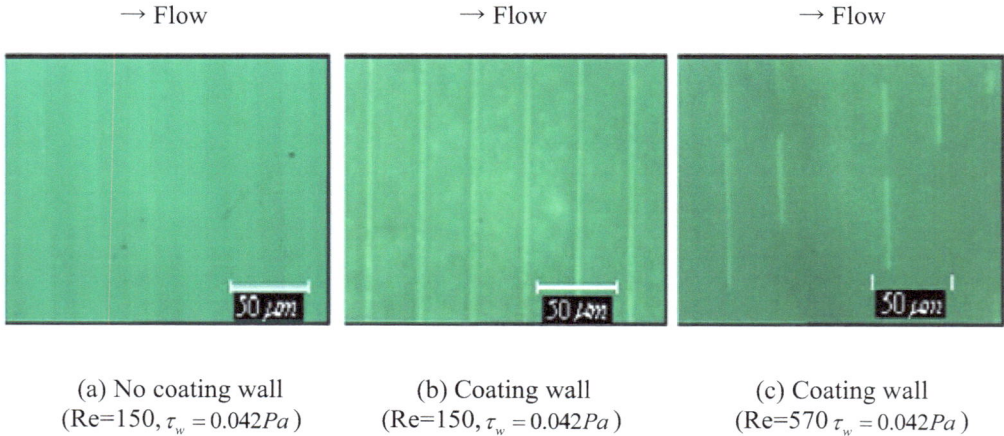

(a) No coating wall   (b) Coating wall   (c) Coating wall
(Re=150, $\tau_w = 0.042Pa$)   (Re=150, $\tau_w = 0.042Pa$)   (Re=570 $\tau_w = 0.042Pa$)

Figure 12: Flow visualization results [5], of air-liquid interface: 5 μ m width grooves.

Because the results agree well with the experimental data in the low Reynolds number range, the prediction for $1.57 \geq \beta \geq 1.17$ can be considered valid for the analytical approach. However, the experimental data come to the solid line according to the increase in the Reynolds number. As described above, this indicates that the gas interface between the duct wall and the liquid is dissolved by the shearing force of the fluid. If these grooves are filled with liquid fluid slip occurs at the wall, as with surfactant solution flow. The identification of this phenomenon was done experimentally using a rectangular duct in a pressure driven flow system where the upper wall of the duct was made of glass with 5 μ m wide grooves etched into the surface. The illuminated surface was observed by using a microscope.

Microphotographs of the wall surface are shown in Fig. **12**. A left-to-right tap water flow is shown in these figures. Because the light source was in the same direction as the microscope and because the pitch of the grooves agreed with that of the white lines, as shown in Fig. **12a**, these white lines are indications of light reflected from the air–tap water interface. They do not appear for the no-coating surface in Fig. **12a** although they do appear in Fig. **12b** and **c**. This means that the air–liquid interface does not exist in the case of a no-coating surface, even if it has fine grooves. In other words, the apparent fluid slip does not occur and the drag reduction does not occur at the surface.

On the other hand, the white lines are observed for the coating wall. However, they abut decrease according to the increase in the Reynolds number, with the interface fading away and no reduction occurring as shown in Fig. **9**. Thus, it is important to maintain the air–liquid interface at the hydrophobic wall surface when we obtain a laminar drag reduction at a high shear rate flow range.

SUMMARY

It is known that an exact solution of the Navier-Stokes equation is obtainable for fully developed laminar flow in a duct. This makes a flow system a useful experimental tool to investigate the surface characteristics of a duct using a small test section to infer the behavior of the larger system. From these considerations, the friction factor of a duct was derived analytically using fluid-slip boundary conditions and was compared with the experimental measurements of a duct with a highly water-repellent wall. Based on the earlier discussion of the measurements of laminar drag reduction, and of the significance of the water repellency of a duct wall with a fractal structure, three types of the drag-reducing wall prototypes were made by wet etching. Notably, experiments measurements on those prototypes suggest that the liquid-slip velocity at the surface depends on the configuration on the surface and on the surface hydrophobicity. Thus, surface structure must be designed to obtain an effective drag reducing wall for a wide laminar flow range in future research.

REFERENCES

[1]   K.Watanabe, Yanuar, K. Okido and H. Mizunuma, (1996) "Drag reduction in Flow through Square and Rectangular Ducts with Highly Water-Repellent Wall", Trans. JSME, **62**-601,102-106.

[2]   K. Watanabe, Yanuar and H. Mizunuma, (1997) "Slip of Newtonian Fluids at Solid Boundary", Trans. JSME, **63**-611.

[3]   K.Watanabe, Yanuar and H. Udagawa, (1999) "Drag reduction of Newtonian fluid in a circular pipe with a highly water-repellent wall", J. Fluid Mech., **381**, 225-238.

[4]   J. P. Hartnett, J. C. Y. Koh, and S. T. McComas, (1962) " A Comparison of Predicted and Measured Friction Factors for Turbulent Flow through Rectangular Ducts, " Trans. ASME, Ser. C, **84**-1, 82-88.
[5]   K. Watanabe, S.Ogata, A. Hirose and A. Kimura, (2007), " Flow Characteristics of Drag Reducing Solid Wall", Int. J. of JSME, J. of Environment and Eng. **2**, 109-114.

Laminar Drag Reduction, 2015, 41-53

CHAPTER 5

Flow Between Two Coaxial Rotating Cylinders

Abstract: The flow pattern between two coaxial rotating cylinders changes from Couette flow to Taylor vortex flow with an increase in the rotation speed. The flow systems described here are Couette flow and laminar Taylor vortex flow. Even though Couette flow is one of the simple shear flows, because it a reasonably good model for certain kinds of friction bearings, it is significant to experimentally clarify the drag reduction phenomenon with fluid slip for the frictional torque acting on a rotating shaft for the practical applications. In the Taylor vortex flow range, the effect of the apparent fluid slip on the vortex formation is examined analytically and experimentally.

Keywords: Drag reduction, two coaxial rotating cylinders, Couette flow, Taylor vortex flow, Göltler vortex, Taylor cell, moment coefficient, highly water-repellent wall, wall shear stress, slip velocity, Non-Newtonian fluid, power-law model.

1. INTRODUCTION

Rheometrical flow systems which have rotating parts, have been the subject of considerable interest concerning the interaction between a liquid and a solid surface. The flow between concentric rotating cylinders is well known as a typical example of Couette flow in the low Reynolds number range. For the case of an outer rotating cylinder and an inner cylinder at rest, although patterns of alternating laminar and turbulent flow are observed [1-3], the stable flow is obtained in a large Reynolds number range. Different types of flow patterns are seen with an increase in the rotation speed as shown in Fig. **1**. First, the drag reduction of the frictional torque of the inner rotating shaft is shown experimentally for Newtonian and non-Newtonian fluids. Then, a flow system with the outer cylinder at rest and the inner cylinder rotating with fluid slip is examined using velocity profile measurements for the Couette flow range. Because a particle image velocimetry (PIV) can be used to measure the velocity adjacent to the wall surface, the experimental data for the slip velocity were measured by means of this technique. When the angular velocity of the inner cylinder is increased, the Couette flow becomes unstable and a secondary steady state is seen as characterized by axisymmetric toroidal vortices, known as Taylor [4] vortex flow as shown in Fig. **2**. The effects of the slip velocity on the wall surface on a on Taylor vortex of an inner rotating cylinder are shown experimentally and analytically.

Keizo Watanabe

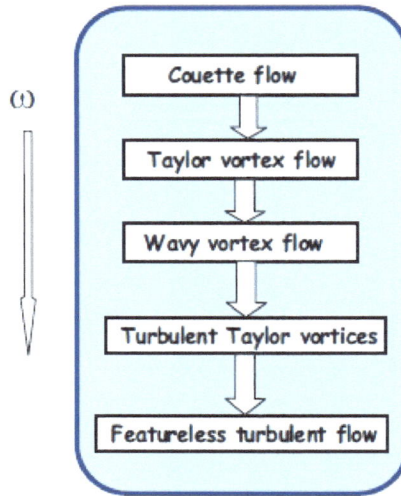

Figure 1: Flow pattern between two coaxial cylinders.

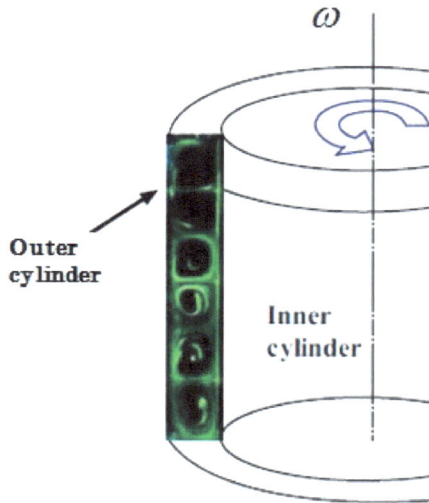

Figure 2: Setup of a Taylor vortex system.

2. COUETTE FLOW

2.1. Frictional Torque

Fig. **3** shows the flow model for this system when the inner cylinder is at rest and the outer cylinder rotates. This flow system is obtained the stable flow as compared with the system that an inner cylinder rotates. Thus, it is frequently used for a rotating viscometer because of the stable measurement result is obtained for the torque acting on the inner cylinder shaft. If the fluid slip occurs on the wall surface of an inner rotating cylinder, the torque [5] acting on the shaft will decrease compared to the case of a no-slip condition at the wall. First, we consider a two-dimensional Newtonian fluid flow with fluid slip at the wall as shown in Fig. **3**. In steady laminar flow, the shear stress $\tau_{r\theta}$ and θ-component of the Navier-Stokes equation are given as follows

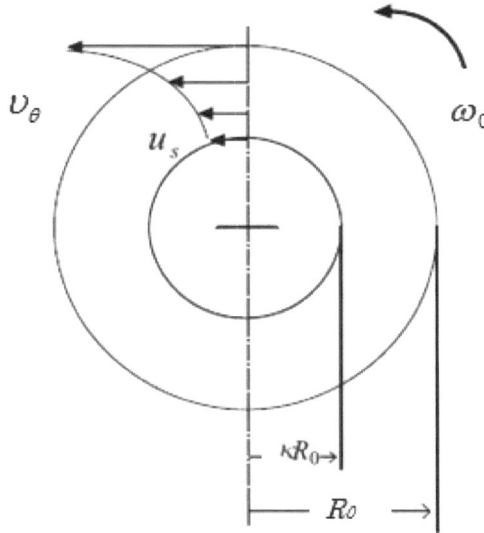

Figure 3: Flow model.

$$\tau_{r\theta} = \mu \left[r \frac{\partial}{\partial r} \left(\frac{\upsilon_\theta}{r} \right) \right] \tag{5-1}$$

$$0 = \frac{d}{dr} \left[\frac{1}{r} \frac{d}{dr} (r\upsilon_\theta) \right] \tag{5-2}$$

where, μ and υ_θ are the viscosity and the velocity of the θ-component, respectively. If the fluid slip occurs at the wall surface of the inner cylinder, the boundary conditions are

$$\left. \begin{array}{l} r = \kappa R_0 \,; \; \upsilon_\theta = u_s \\ r = R_0 \,; \; \upsilon_\theta = \omega_0 R_0 \end{array} \right\} \tag{5-3}$$

where, u_s and ω_0 are the slip velocity at the wall and the angular velocity of the outer cylinder, respectively. The velocity profile υ_θ is calculated using Eq. (5-2) under the boundary conditions in Eq. (5-3).

$$\upsilon_\theta = \frac{\kappa R_0}{r} u_s + \frac{R_0 \omega - \kappa u_s}{(1 - \kappa^2) R_0 r} (r^2 - \kappa^2 R_0^2) \tag{5-4}$$

By substituting Eq. (5-4) into Eq. (5-1), we can obtain the frictional torque T required to turn the outer shaft.

$$T = 4\pi\mu R_0^2 L \left(\omega_0 \kappa - u_s \frac{1}{R_0} \right) \left(\frac{\kappa}{1 - \kappa^2} \right) \tag{5-5}$$

Thus the slip velocity u_s is related to the torque T by

$$u_s = R_0 \kappa \omega_0 - \frac{T}{4\pi\mu R_0 L} \frac{(1 - \kappa^2)}{\kappa} \tag{5-6}$$

By substituting the measured value of the torque T into Eq. (5-6), we can calculate the slip velocity u_s. On the other hand, Eq. (5-6) can be modified by using $\tau_{r\theta}|_{r=\kappa R_0} = \beta u_s$ which is obtained by the assumption that Eq. (1) in Chapter 1 is satisfied at the wall, to form

$$u_s = \frac{2\mu R_0 \kappa \omega_0}{\beta \kappa R_0 (1 - \kappa^2) + 2\mu} \tag{5-7}$$

where, β is the sliding constant.

The coefficient of torque acting on the inner cylinder C_m is obtained as

$$C_m = \frac{16\kappa^2}{(\kappa + 1)^4} \left[1 - \frac{2}{\frac{\beta R_0}{\mu} \kappa (1 - \kappa^2) + 2} \right] \Bigg/ R_\omega \tag{5-8}$$

where, C_m is defined as $C_m = T / 2\pi\rho R_m^4 \omega_0^2 L$. R_ω is the Reynolds number as defined by $R_\omega = R_0 (1 - \kappa) R_m \omega_0 / \nu$ where R_m is the mean radius as given by

$R_m = R_0(1+\kappa)/2$. When no fluid slip occurs, we obtain the following equation by substituting $\beta \to \infty$ into Eq. (5-8)

$$C_m = \frac{16\kappa^2}{(\kappa+1)^4} \Big/ R_\omega = \frac{16(1-\delta/R_0)^2}{2^4(1-\delta/2R_0)^4} \Big/ R_\omega = \frac{(1-\delta/R_0)^2}{\{1-\delta/R_0+(\delta/2R_0)^2\}^2} \Big/ R_\omega \tag{5-9}$$

where, δ is the gap between the outer and inner cylinders. In the case of $(\delta/R_0)^2 \cong 0$, we obtain $C_m = 1/R_\omega$. For a small gap, the term $(\delta/R_0)^2$ is very small, such as just shown above, and we have $C_m = 1/R_\omega$.

In laminar non-Newtonian fluid flow, if we assume a power-law model, the shear stress and θ-component of the equation of motion are given by the following equations, respectively,

$$\tau_{r\theta} = K\left[r\frac{\partial}{\partial r}\left(\frac{v_\theta}{r}\right)\right]^n \tag{5-10}$$

$$0 = \frac{1}{r^2}\frac{\partial}{\partial r}\left(r^2\tau_{r\theta}\right) \tag{5-11}$$

where, K and n are a consistency index and an exponent in a power law model, respectively. The slip velocity u_s is calculated using Eqs. (5-10) and (5-11) under the fluid slip boundary conditions shown in Eq. (5-3) as follows

$$u_s = \kappa\omega_0 R_0 - \frac{1}{2}n\left(\kappa R_0\right)^{1-(2/n)}\left(1-\kappa^{2/n}\right)\left(\frac{T}{2\pi LK}\right)^{1/n} \tag{5-12}$$

On the other hand, for no slip fluid slip C_m is given by the following equation by applying $(\delta/R_0)^2 \equiv 0$.

$$C_m = \frac{2^{2n+2}\kappa^2\left(1-\kappa^2\right)}{(1+\kappa)^4 n^n\left(1-\kappa^{2/n}\right)^n R_\omega^*} \tag{5-13}$$

where, R_ω^* is a modified power law Reynolds number defined by $R_\omega^* = \left\{R_0(1-\kappa)^2\omega_0^{2-n}/2^{2-n}K\right\}$. If we substitute $n=1$ into Eqs. (5-12) and (5-13), they agree with analytical results for a Newtonian fluid.

The experimental results for the moment coefficient values of glycerin solutions are shown in Fig. **4a** and **b**. It was the most large error that the reported the torque measurement at R_ω=10 and κ=0.932 in 60% glycerin solutions is best estimate of the result, and with 99% confidence, true value is believed to lie within 1.85% of the estimated result. In these figures, the solid denotes $C_m = 1/R_\omega$ for the no-slip slip boundary condition. It is interesting to note that at a highly water-repellent or hydrophobic wall, the drag reduction effect increases with increases in the concentration and aspect ratio although the experimental data for Teflon (PTFE polytetrafluoroethylene) wall agree with the solid line. On the other hand, this flow system has an advantage for the measurement of the slip velocity because we can directly examined the slip velocity using the torque measurements as shown in Eq. (5-6). The relationship between the slip velocity and the shear stress is shown in Fig. **5a** and **b**. Clearly, in Fig. **5a** for a glycerin solution that is a Newtonian fluid the relationship between the wall shear stress and the slip velocity is linearity. Thus the gradient of the straight line gives the sliding constant β as shown by Eq. (2-4) in Chapter 2 [6]. The dotted lines are the calculated results of Eq. 8 using the numerical value of β obtained in these figures. As a matter of course experimental data agree well with Eq. (5-8). Fig. **5b** shows the experimental data for polymer solutions, which are non-Newtonian fluids. The experimental results represent a non-linear behavior as assumed by Eq. (3-4) in Chapter 3.

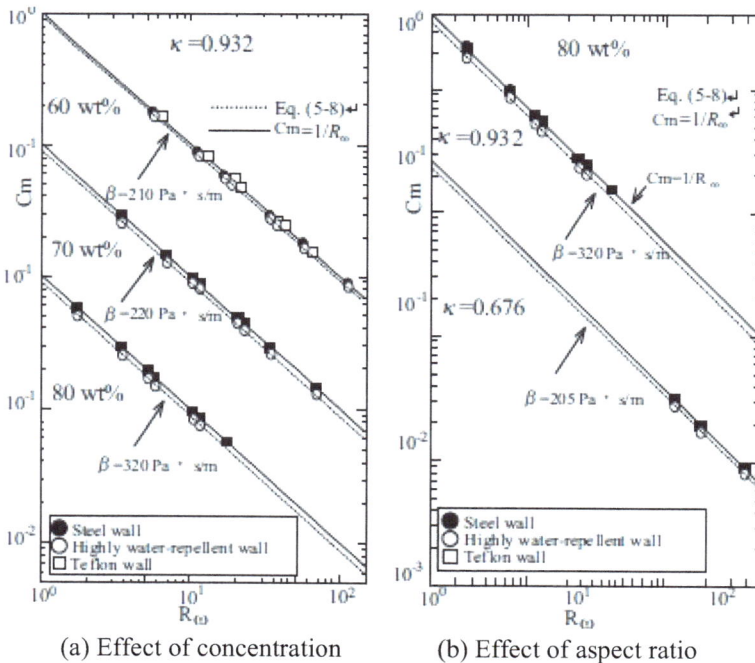

(a) Effect of concentration   (b) Effect of aspect ratio

Figure 4: Moment coefficients of glycerin solutions.

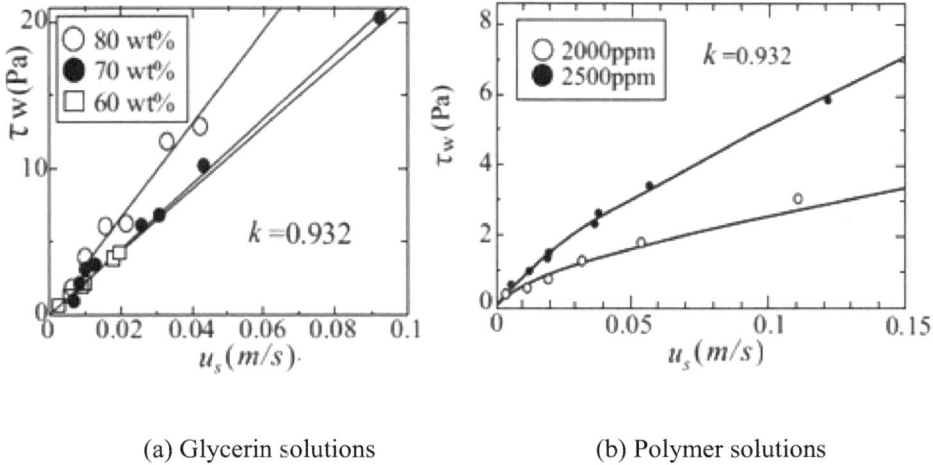

(a) Glycerin solutions　　　　(b) Polymer solutions

Figure 5: Slip velocities estimated from measured torque values.

2.2. Velocity Profile

A problem remains with measuring the velocity by means of an instrumentation probe. We cannot measure the value in the vicinity of the wall because of the size of the measurement probe. PIV is an optical method for the flow visualization and the experimental data for the velocity in the wall vicinity are comparatively easy to obtain from a visualization photograph. Thus this is an advantageous method for determining the slip velocity at the wall. Fluid slip does not occur at the inner aluminum cylinder wall as well known in general, and the experimental data fit the Eq. (5-4) values obtained by substituting $u_s = 0$. Velocity profiles [7] for a hydrophobic wall are shown in Fig. 6. In these figures, δ is the clearance between the inner and the outer cylinders and equals to $(R_0 - R_i)$. The effects of the aspect ratio and of glycerin concentration on the velocity profile are shown in Fig. **6a** and **b**, respectively. Fig. **6a** and **b** show that the fluid velocity does not agree with that of the inner wall velocity, that is the fluid slip occurs at the wall. The data in Fig. **6a** show that the slip velocity has a maximum value when the radius ratio $\kappa = 0.839$, while in Fig. **6b**, the slip velocity increases slightly with the glycerin concentration increase. The non-dimensional slip velocity ratio $u_s / R_i \omega_i$ is equal to about 0.17 for $\kappa = 0.839$ in an 80% glycerin solutions, and the slip velocity is weakly dependent on the radius ratio in this experiment range. It can be seen that the sliding constant β is determined by the slip velocity and the solid lines obtained by substituting β into Eq.(5-4) agree well with the experimental data as shown in Fig. **6**.

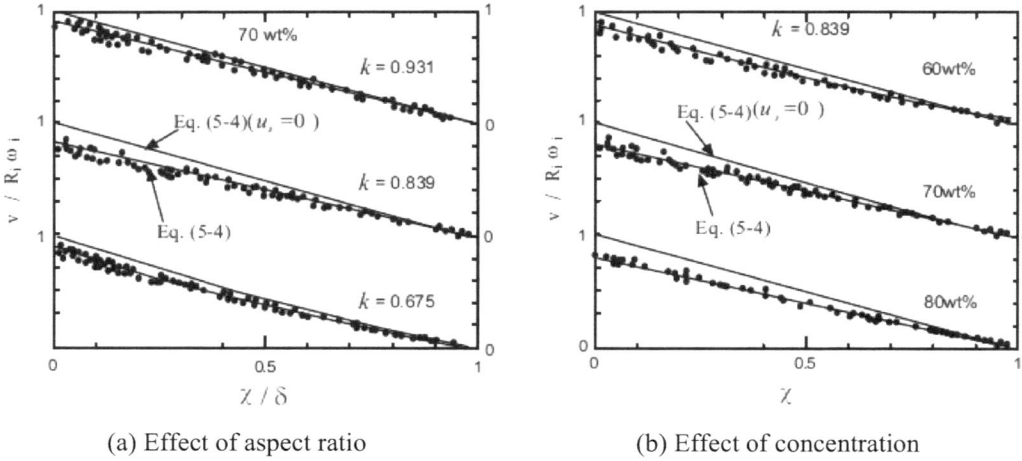

(a) Effect of aspect ratio   (b) Effect of concentration

Figure 6: Velocity profiles of glycerin solutions.

3. TAYLOR VORTEX FLOW

As is well known, Couette flow changes into Taylor vortex flow when the angular velocity of the inner cylinder is increased. The flow shows a stable secondary flow pattern in which large toroidal vortices form in the flow. Taylor showed that Taylor vortices formed when the Taylor number of the flow exceeded a critical value. Because the rotating speed of the inner cylinder is increased from zero to the constant angular velocity in the actual experimental apparatus, the vortices are formed by the initial deployment of a Göltler vortex. It is possible to clarify the formation process using flow visualization measurements of the vortex. The experimental results [8] for smooth wall are shown in Fig. **7a-f**. In these figures, it takes 10 sec. to reach a constant angular velocity, and the numerical values indicate the elapsed time for the rotation of the inner cylinder from the beginning of rotation. After the inner cylinder has rotated at a constant angular velocity for 10 sec., some small vortices are generated from the surface of the inner cylinder as shown in Fig. **7a**. They change to Göltler vortices, and six stable six vortices are formed as shown in Fig. **7f** after about 3 h. The number of Taylor vortices equal to the integer number of the aspect ratio Γ of the experimental apparatus as shown by Taylor. Therefore, six vortices appear in this case as shown in Fig. **7f** because it is $\Gamma = 6.15$. In general, the stability of the vortex depends on the Reynolds number, and the axial symmetry will be maintained in the laminar flow range. Although many experimental analytical studies have considered the flow range, very few have investigated a flow with fluid slip on a rotating inner cylinder.

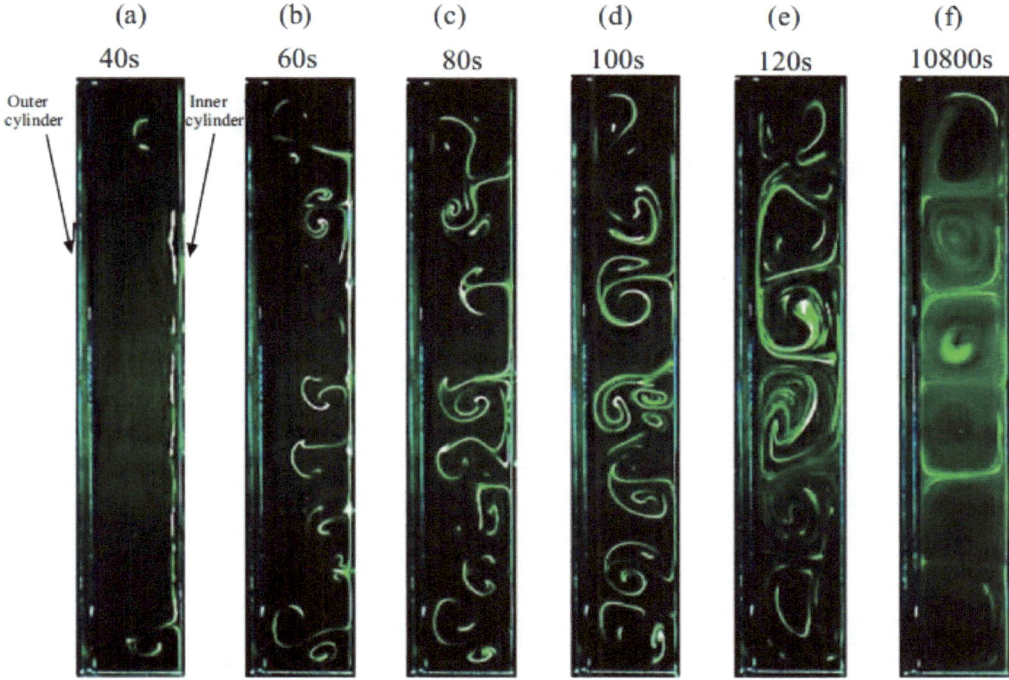

Figure 7: Formation process for Taylor vortex (Re=500) for a smooth wall cylinder.

We can examine the effect of fluid slip at the inner cylinder on the Taylor vortex by using the Navier-Stokes equations and the continuity equation. It is assumed as the flow of an incompressible fluid with kinematic viscosity ν between two cylinders of height L, and with inner radius R_i and outer radius R_o. Fig. **8** shows the coordinate system for the flow field. The governing equation is set up using Navier-Stokes equations and the continuity equation for the axis symmetry flow. They are given as the flowing equations,

$$u\frac{\partial u}{\partial r} + w\frac{\partial u}{\partial z} - \frac{v^2}{r} = -\frac{\partial p}{\partial r} + \frac{1}{R_\omega}\left(\frac{\partial^2 u}{\partial r^2} + \frac{1}{r}\frac{\partial u}{\partial r} - \frac{u}{r^2} + \frac{\partial^2 u}{\partial z^2}\right) \tag{5-14}$$

$$u\frac{\partial v}{\partial r} + w\frac{\partial v}{\partial z} + \frac{uv}{r} = \frac{1}{R_\omega}\left(\frac{\partial^2 v}{\partial r^2} + \frac{1}{r}\frac{\partial v}{\partial r} - \frac{v}{r^2} + \frac{\partial^2 v}{\partial z^2}\right) \tag{5-15}$$

$$u\frac{\partial w}{\partial r} + w\frac{\partial w}{\partial z} = -\frac{\partial p}{\partial z} + \frac{1}{R_\omega}\left(\frac{\partial^2 w}{\partial r^2} + \frac{1}{r}\frac{\partial w}{\partial r} + \frac{\partial^2 w}{\partial z^2}\right) \tag{5-16}$$

$$\frac{\partial u}{\partial r} + \frac{\partial w}{\partial z} + \frac{u}{r} = 0 \tag{5-17}$$

where, u, υ and w are the velocity of r, θ and z the directions, respectively. p is the pressure and where the Reynolds number R_ω, is defined as $R_\omega = R_i \omega_i \delta / v$. Here, δ is the gap given as $(R_0 - R_i)$.

On the other hand, the velocity components u and w are obtained by using the Stokes stream function, Ψ,

$$u = \frac{1}{r} \frac{\partial \Psi}{\partial z} , \quad w = -\frac{1}{r} \frac{\partial \Psi}{\partial r} \tag{5-18}$$

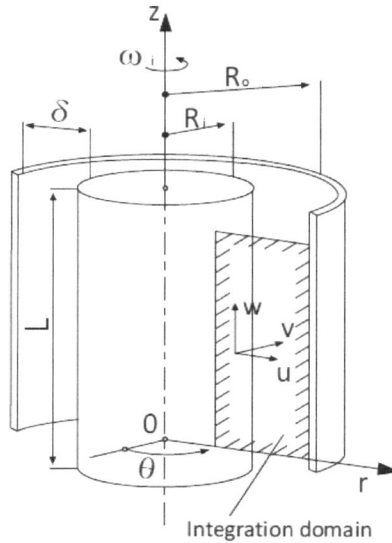

Figure 8: Coordinate system for flow field.

The boundary conditions for no-slip and slip flows are given as follows;

No-slip at smooth wall

$$
\begin{aligned}
u = w = 0, \upsilon = R_i \omega_i &&& r = R_i \\
u = \upsilon = w = 0 &&& r = R_o, \quad and\ z = 0 \\
\partial u / \partial z = \partial \upsilon / \partial z, w = 0 &&& z = L/2
\end{aligned}
$$

Fluid slip for hydrophobic wall

$$u = 0, \upsilon = R_i\omega_i - V_s, w = -W_s \qquad r = R_i$$
$$u = \upsilon = w = 0 \qquad\qquad\qquad r = R_o, and\ z = 0$$
$$\partial u/\partial z = \partial \upsilon/\partial z, w = 0 \qquad\quad z = L/2$$

where, V_s and W_s are the slip velocities in the tangential and axial

directions on the wall of the inner cylinder, respectively. These slip velocities were obtained by applying

are given using by Eq. (1-1) in Chapter 1.

$$\tau_{r\theta}|_{r=R_i} = \beta V_s, \ \tau_{rz}|_{r=R_i} = \beta W_s \qquad\qquad\qquad (5\text{-}19)$$

Fig. **9** shows the stream lines obtained from the numerical work. In these figures, X^* is the non-dimensional radius from the outer cylinder wall, $\dfrac{(R_0 - r)}{\delta} = \dfrac{(R_0 - r)}{(R_0 - R_i)}$. Z^* is the non-dimensional height, $\left(\dfrac{z}{\delta}\right) = \left(\dfrac{z}{R_0 - R_i}\right)$, and three Taylor cells are shown in half of the cylinder length $z/\delta = 3.075$.

Taylor pointed out that the number of cells is approximately equal to the aspect ratio. Thus, six Taylor cells exist in the flow between two coaxial cylinders. In addition, the fluid moves from the outer cylinder to the inner cylinder at the bottom of the annular space. The result agrees with the previous research [3]. When fluid slip occurs at the inner cylinder wall, the computation result shows that the number and flow direction of the Taylor cells are the same as in the case of the no fluid slip. However, comparing the size of the region where the absolute value of the stream function is high, the radius and axial velocity components for the fluid slip wall are smaller than those for a no-fluid slip wall. These results agree qualitatively with the experimental flow visualization results for $100 < R_\omega < 700$.

Fig. **10** shows the effect of the viscosity on the height of Taylor cell-1, which touches the bottom of the cylinder at the highly water-repellent wall and smooth wall. In Fig. **10**, h^* is the non dimensional height, in which is expressed as (h/δ). Regardless of the existence of fluid slip, the height of cell does not depend on the Reynolds number R_ω in the range of not smaller than 700. However, the height of the cell increases with an increase in the viscosity for $100 < R_{\omega} < 700$. The results revealed that the height depends on not only on R_{ω} but also on the non-dimensional parameters including the viscosity. In particular, the effect of fluid slip becomes remarkable as the viscosity increases. This tendency agrees well with the measurement results for the measurement of the velocity profile of the Couette flow. On the other hand, it has been clarified that an interesting phenomenon occurs in the formation process of the vortex of polymer solutions [9].

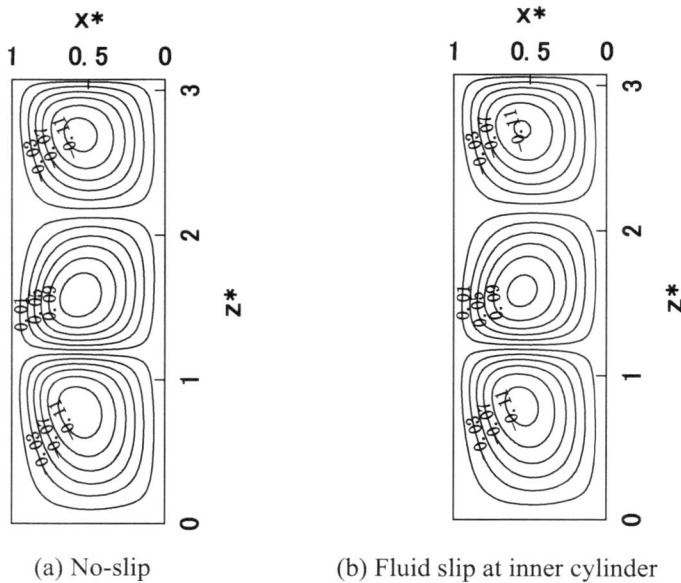

(a) No-slip   (b) Fluid slip at inner cylinder

Figure 9: Analytical results for Taylor vortex (Re=300).

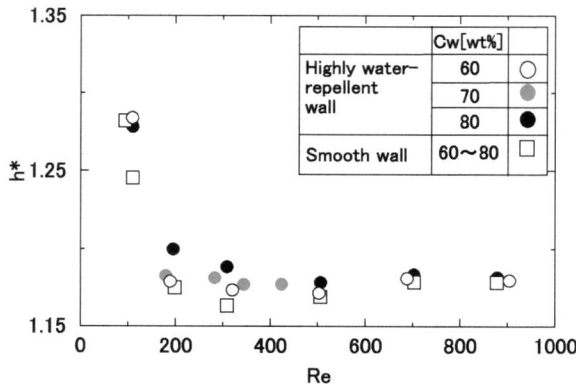

Figure 10: Variation of Taylor cell-1 heights with concentration of glycerin solution ($\Gamma = 6.15$).

SUMMARY

Because the flow between two coaxial rotating cylinders changes from a simple shear flow to a complex Taylor vortex flow with a change in the rotation speed, we were able to conveniently obtain much experimental data related to fluid slip. The torque measurements showed that a laminar drag reduction could be achieved in a laminar Couette flow with a highly water repellent wall, and the slip at the wall was clearly identified from PIV measurements of the velocity in a flow field including the vicinity of a rotating inner cylinder surface.

The maximum slip was found to occur when the radius ratio κ was equal to 0.839 in this experiment, and the maximum slip velocity at the surface was 18% of the rotating cylinder wall velocity. The slip velocity was affected by an increase in the viscosity of an increase. The flow of the Taylor cells become weak, and the slip velocity in the tangential direction decreased the pressure difference between the two cylinders. The slip velocity in the axial direction caused a difference in the size of the Taylor cells.

REFERENCES

[1]   Coles, D.,(1965), "Transition in Circular Couette Flow," J. Fluid Mech.,**21**, Part 3, 385-425
[2]   Van Atta, C., (1966), "Measurements in Spiral Turbulence," J. Fluid Mech., **25,** Part 3, 495-512
[3]   Benjamin, T. B., (1979)," Bifurcation Phenomena in Steady Flows of a Viscous Liquid, II . Theory," Philos. Trans. Roy. Soc. London, **A359**, 27-45.
[4]   Taylor, G. I., (1923), "Stability of a viscous liquid contained between two rotating cylinders", Phil. Trans. Royal Soc. **A223**, 289-343.
[5]   Watanabe K., and T. Akino, (1999) "Drag Reduction in Laminar Flow Between Two Vertical Coaxial Cylinders", Trans. of ASME, Journal of Fluids Engineering, **121**, pp.541-546.
[6]   Watanabe, K., Yanuar and Y. Udagawa, (1999) "Drag reduction of Newtonian fluid in a circular pipe with a highly water-repellent wall," J. Fluid Mech., **381**, 225-238.
[7]   Watanabe, K., *et al.*,(2003),"Flow between Two Coaxial Rotating Cylinders with a Highly Water-Repellent Wall". AIChE J. **49,** 8, 1956-1963.
[8]   Watanabe K., T. Takayama and S. Ogata, (2004) "Instability of Surfactant Solution Flow in a Taylor Cell," J. of Visualization, **7, 4**, p.1-9.
[9]   Watanabe K., S. Sumio and S. Ogata, (2006), "Formation of Taylor Vortex Flow in Polymer Solutions," Trans. ASME, J. Fluid Eng., 128, 95-100.

Chapter 6: *Flow Near a Rotating Disk*

Laminar Drag Reduction, 2015, 55-70

Flow Near a Rotating Disk

Abstract: Flow near a rotating disk with constant angular velocity is a typical example of a three dimensional boundary-layer flows. In the case where the flow around a disk is rotating in a housing, *i.e.*, when the rotating disk is enclosed, we can apply the flow model of a turbo-machinery impeller in a casing to estimate the frictional torque. Thus, many studies have been performed to clarify the characteristics of such flow. In this chapter, experiments were carried out to measure the velocity profile and frictional torque acting on a rotating disk with fluid slip in a Newtonian fluid filled chamber. A disk with a highly water-repellent wall gave rise to the drag reduction phenomenon when the moment coefficient was in the laminar flow range. Analysis of the moment coefficient by using momentum integral equations with the fluid slip boundary condition revealed results that qualitatively agreed with the experimental results.

Keywords: Drag reduction, rotating disk, frictional torque, boundary layer flow, Couette flow, highly water-repellent wall, fluid slip, momentum integral equation, moment coefficient, sliding constant, flow angle.

1. INTRODUCTION

From the viewpoint of fluid engineering the flow near a flat disk that rotates about an axis perpendicular to its plane with a uniform angular velocity in a fluids is classified into the following categories: a rotating disk in an infinity fluid, a free disk and rotating disk in a chamber-filled fluid; and an enclosed rotating disk. In the first case, the thin layer on the disk surface is regarded as fully three-dimensional boundary layer. The flow characteristics are solved by using boundary-layer theory, and the frictional torque acting on the disk surface is calculated from the results. This first case was initially analyzed as an example of an exact solution of the Navier-Stokes equations for a laminar boundary layer by Von Kármán [1] who also analyzed the frictional torque in the turbulent boundary-layer flow range by using momentum integral equations. These analytical results for the moment coefficient quantitatively agree with the experimental results.

A turbo-machinery impeller with a disk, which is a centrifugal pump or a turbine, is rotated in a casing with a narrow gap. The frictional torque acting on an enclosed rotating disk surface is an important factors that affects the performance of an impeller in turbo-machinery a large class of machine that has become ubiquitous in the developed world. The flow near an enclosed disk is closely related to flow in an impeller and we can apply to examine the friction loss of impeller using the analytical or experimental results [2, 3]. In general, the friction loss for the rotating disk of a centrifugal pump impeller is about 5%-15% of the total efficiency. From the perspective energy conservation, an important issue is to obtain the drag reduction on the frictional torque of an enclosed rotating disk. If successful, turbo-machinery performance could be elevated. Studies on turbulent drag reduction in a rotating disk are conducted by using drag-reducing additives such as high molecular weight polymers [4, 5] or surfactants [6] to improve the performance of turbo-machinery. Although these drag-reducing additives reduce the power loss of turbo-machinery, the difficulty of sustaining this effect remains because the solutions undergo degradation and environmental load increases, in addition we cannot apply the drag reducing additives to the laminar flow because turbulence is not in the flow range. In this chapter, the pattern of flow near an enclosed rotating disk is experimentally clarified with highly water-repellent wall in Newtonian fluids. and after the frictional moment is analyzed using the boundary conditions which the slip velocity of fluids existed at the disk wall. The analytical results of the moment coefficient obtained using the fluid slip boundary conditions are compared with the experimental results.

2. CASE OF A ROTATING DISK SEPARATED FROM HOUSING BY VERY SMALL GAPS

The velocity profiles become particularly simple when the flow is laminar and when the gap between the rotating disk and the housing is very small. As shown in Fig. **1**, if the gap, s, is smaller than the boundary layer thickness, we can obtain the frictional torque using the Couette flow model.

The Navier-Stokes equations is then given

$$\frac{\partial^2 \upsilon_\theta}{\partial z^2} = 0$$

The boundary conditions for slip flow are,

$$z = 0 \; : \; \upsilon_\theta = \left(r\omega - u_{s\theta}\right)$$

$$z = s \; : \; \upsilon_\theta = 0$$

where, $u_{s\theta}$ is the slip velocity.

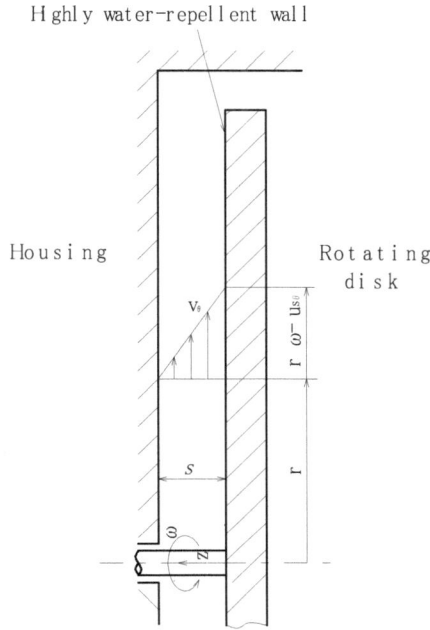

Figure 1: Couette flow model.

Thus, the velocity profile is obtained as follows

$$\upsilon_\theta = \left(r\omega - u_{s\theta}\right)\left(1 - \frac{z}{s}\right)$$

By using Eq. (1-1) in Chapter 1, the slip velocity, $u_{s\theta}$ is obtained as $u_{s\theta} = \dfrac{\mu r\omega}{\beta s + \mu}$

and the frictional torque on one side of a disk is given by

$$M = -\int_0^a 2\pi\tau_\theta\, r^2\, dr = \frac{\pi\mu\omega\beta\, a^4}{2(\beta s + \mu)}.$$

The moment coefficient becomes

$$C_m = \frac{2M}{\rho\omega^2 a^5} = \frac{\pi}{\mathrm{Re}}\left(\frac{a}{s}\right)\left[\frac{1}{1 + (\mu/\beta s)}\right] \qquad (6\text{-}1)$$

If fluid slip does not occur, by inserting $\beta \to \infty$ into Eq. (6-1) yields

$$C_m = \frac{\pi}{\mathrm{Re}}\left(\frac{a}{s}\right)$$

Eq. (6-1) is plotted as the solid lines in Fig. **2** for $(s/a) = 0.011$ at various values of β. The dotted line in the figure is semi-theoretical result of Daily & Nece. Eq.(6-1) may be used to predict the drag reduction for fluid slip in the range of $\mathrm{Re} < 10^4$.

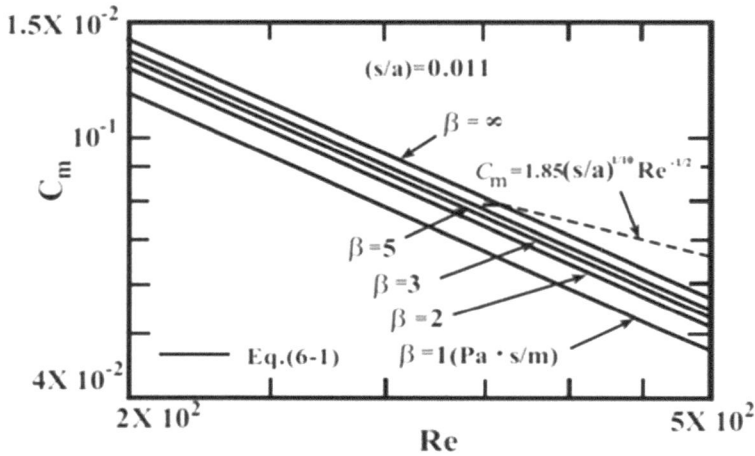

Figure 2: Moment coefficients for very small gaps.

3. CASE OF A DISK IN LARGER GAPS

3.1. Velocity Profile

Depending on the Reynolds number as well as the size of the gaps between the rotating disk and housing surface, the flow pattern in the gaps considerably differs from that of the simple scheme described above. In general, the flow pattern comprises fluid core already in the gap and a boundary layer formed on the disk and housing. The fluid core has no radial velocity component and rotates at about half the angular velocity of the disk. Fig. **3** shows the flow pattern in the gap. In the figure, K equals the ratio of the angular velocity of the fluid core to that of the disk. The fluid in the boundary layer on the rotating disk is centrifuged outwards. Fig. **4** shows streak lines of the near disk surface flow visualized using an oil film technique [7]. The inclinations of these streak lines denote the direction of the wall shear stress.

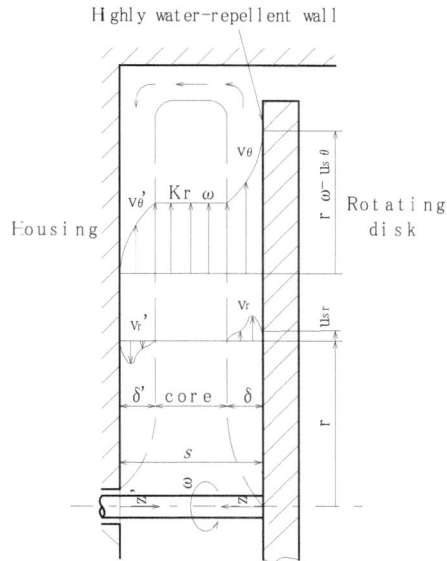

Figure 3: Flow pattern in the gap.

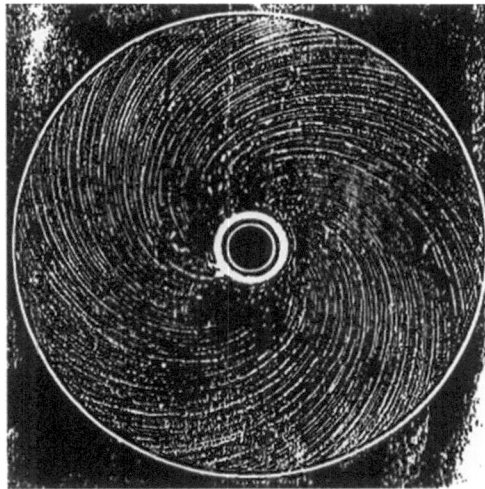

Figure 4: Streak lines on the disk.

The inclination becomes

$$\tan \varphi = \left| \frac{\tau_r}{\tau_\theta} \right| = -\left(\frac{\partial \upsilon_r / \partial z}{\partial \upsilon_\theta / \partial z} \right)_{z=0} \qquad (6\text{-}2)$$

where, τ_r and τ_θ correspond with the wall shear stress for the streak lines as shown in Fig. **4**, respectively. In the case of the laminar boundary-layer flow of a free disk, ϕ is obtained as $\phi = 39.6°$ by using the function for velocity calculated by Cochran [8]. Although the calculation was obtained for an infinite disk, we can utilize the result for the flow near a rotating disk in a housing. Because ϕ corresponds to the flow direction of the boundary layer if the disk wall is imagined at rest, we can measure the value by using the experimental results for the velocity profile of the boundary layer. PIV is a widely used method for measuring the velocity of the flow near a rotating disk. As shown in Fig. **5**, when the solid line that corresponds with the path line taken by the tracer particle, the flow angle is directly obtained by measuring the direction.

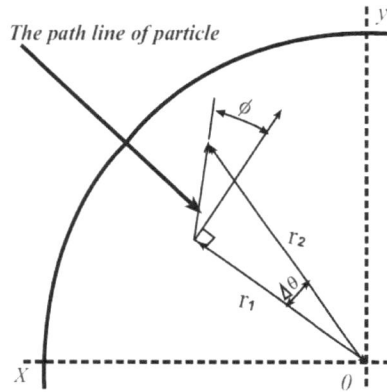

Figure 5: Flow angle.

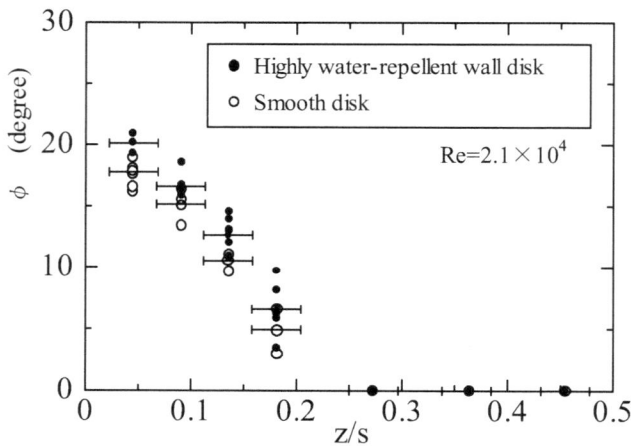

Figure 6: Relationship between flow angle and gap.

Fig. **6** shows the experimental results for the flow angle of the smooth and highly water-repellent wall disks with fluid slip. The flow turns outward with respect to the circumferential direction in the case of slip flow relative to the flow in the case of the smooth disk with no slip. As z increases ϕ decreases and we obtain $\phi = 0$ at $(z/\delta) \cong 0.25$. Furthermore, that the fluid core which rotates with a constant angular velocity because the radial velocity becomes zero at this range exists in the gap is noteworthy. On the basis of these results, the velocity profile in the gap of the housing can be obtained. Fig. **7** shows the velocity profiles around an enclosed rotating disk at Re = 2.1 x 10^4. Experiments were carried out to measure the velocity profile by means of PIV for the experimental apparatus with a diameter d = 180 mm and a clearance s = 10 mm between the disk and the housing wall. A fluid in the range of $z/s > 0.2$, which has no radial velocity profiles, has only tangential velocity. Fig. **8** shows the experimental results of K at $Re = 2.1$ x 10^4 and 4.3 x 10^4. The results of K the ratio of the angular velocity of the fluid to that of the disk are comparable for highly water-repellent wall disk and the smooth disk. The values of K are 0.35~0.4 with an average of 0.38, and K increases slightly a (r/a) increases. In Fig. **7**, Eqs. (6-3) and (6-4) are fitted to the velocity profiles obtained by Daily and Nece [3] under the no-nslip condition. They are written as following equations,

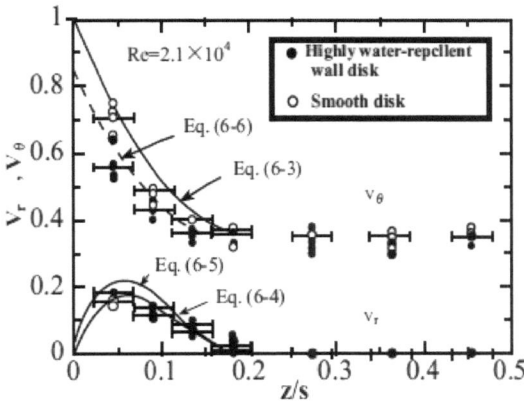

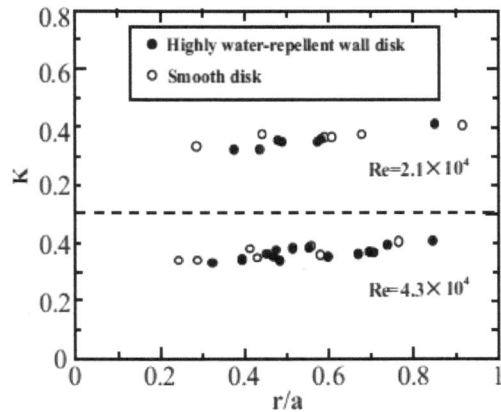

Figure 7: Velocity profile of an enclosed rotating disk. **Figure 8:** K *versus* (r/a).

Radial velocity profile:

$$v_r = \alpha \omega r \left\{ 2\left(\frac{z}{\delta}\right) - \left(\frac{z}{\delta}\right)^2 \right\} \left\{ 1 - \left(\frac{z}{\delta}\right) \right\}^2 \qquad (6\text{-}3)$$

Tangential velocity profile:

$$\upsilon_\theta = \omega r \left[1 + (1-K)\left\{\left(\frac{z}{\delta}\right)^2 - 2\left(\frac{z}{\delta}\right)\right\}\right] \qquad (6\text{-}4)$$

Here ω and δ are the angular velocity and boundary layer thickness, respectively. In Fig. **7**, the results for the smooth wall disk with no fluid slip agree well with Eqs. (6-3) and (6-4). In the case of the highly water-repellent wall disk with fluid slip, the radial velocity increases and the tangential velocity decreases relative to those for the smooth wall disk in the range of $z/s < 0.2$. In other words, the slip velocity may be assumed to occur at the disk wall, because the slip velocity has been obtained by measuring the velocity profiles for a circular pipe with highly water-repellent walls in the laminar flow range as described in Chapter 3. K ratio the angular velocity of the fluid core to that of disk surface, almost shows good agreement between the two disks Fig. **8**. The tracer path lines of the highly water-repellent wall disk were also turned radically outwards relative to those of the smooth disk as shown in Fig. **6**. In other words, the tangential component of the shear stress is decreased. To obtain analytical results for the frictional torque acting on the rotating disk surface by using the momentum integral equations, the velocity profile of the boundary-layer on the disk must be determined. Thus, by drawing upon these experimental results, the velocity profiles of highly water-repellent wall disk [9] are determined as follows,

Radial velocity profile:

$$\upsilon_r = \alpha\omega r \left\{ 2\left(\frac{z}{\delta}\right) - \left(\frac{z}{\delta}\right)^2 + C_1\right\}\left\{1 - \left(\frac{z}{\delta}\right)\right\}^2 \qquad (6\text{-}5)$$

Tangential velocity profile:

$$\upsilon_\theta = \omega r \left[1 + (1-K)\left\{\left(\frac{z}{\delta}\right)^2 - 2\left(\frac{z}{\delta}\right) + C_2\left(\left(\frac{z}{\delta}\right)-1\right)\right\}\right] \qquad (6\text{-}6)$$

where, C_1 and C_2 are unknown parameters. The boundary conditions are given as fpllows;

Radial direction:

$z=0$: $\upsilon_r=u_{sr}=\alpha\omega rC_1$

$z=\delta$: $\upsilon_r= 0$

Tangential direction:

$z=0$: $\upsilon_\theta=\omega r-u_{s\theta}=\omega r-(1-K)\omega rC_2$

$z=\delta$: $\upsilon_\theta= K\omega r$

Thus, these unknown parameters C_1 and C_2 can be determined if the slip velocity on the wall is given as Eq.(1-1) in Chapter 1；

$$\upsilon_r\big|_{z=0}=u_{sr} = 2\mu\alpha\omega r / \left(\beta\delta + \mu\right) \tag{6-7}$$

$$(\omega r - \upsilon_\theta\big|_{z=0}) = u_{s\theta} = 2\mu(1- K)\omega r / \left(\beta\delta + 2\mu\right) \tag{6-8}$$

An example of the determined velocity profiles of Eqs. (6-5) and (6-6) are shown in Fig. **7** by using the dotted lines. The parameters are set to the following values: $K = 0.38$, $\alpha = 0.7$, $C_1 = 0.05$ and $C_2 = 0.25$. They are determined by comparison with the experimental data

3.2. Frictional Torque

(1) Experimental Approach

The frictional torques acting on the bearing and the thickness of the rotating disk are measured in the case of a conventional experimental apparatus for an enclosed rotating disk. Thus a highly accurate experimental apparatus must be fabricated to measure the frictional torque acting on the rotating disk surface only for the laminar drag reduction. In relation to the frictional torque of the bearing, we will be able to omit the frictional torque by setting the torque pick-up on the mounting attachment of the rotating disk as reported by Daily and Nece. However, an apparatus to eliminate the frictional torque that affects the thickness of the rotating disk needs to be developed.

Fig. **9a** and **b** show the experimental apparatus [9] for measuring the torque on only one side of the rotating disk surface and a detailed view of the torque measurement device, respectively. In Fig. **9a**, two D-C motors ⑭ with a speed range of 50-2000 rpm served as the drive units, and were rotated at the same speed with the same direction of rotation. The two shafts speeds were measured by means of a control unit attached to the D-C motors. The first motor was directly coupled to shaft 1 ⑧, to which interchangeable disks ② could be attached, and the other motor was directly coupled to the shaft 2 ⑨, to which support disk ③ was attached. A bronze support disk ③ was used it covered the one side and the tip of the tested disk. The axial clearance between the disk and the support disk was 3 mm, as was the radial clearance. The support disk was rotated at the same speed and direction as the tested disk by controlling the two D-C motors. Thus, we were able to measure the torque acting on only one side of the disk surface by removing the effect of disk thickness on the measurement value.

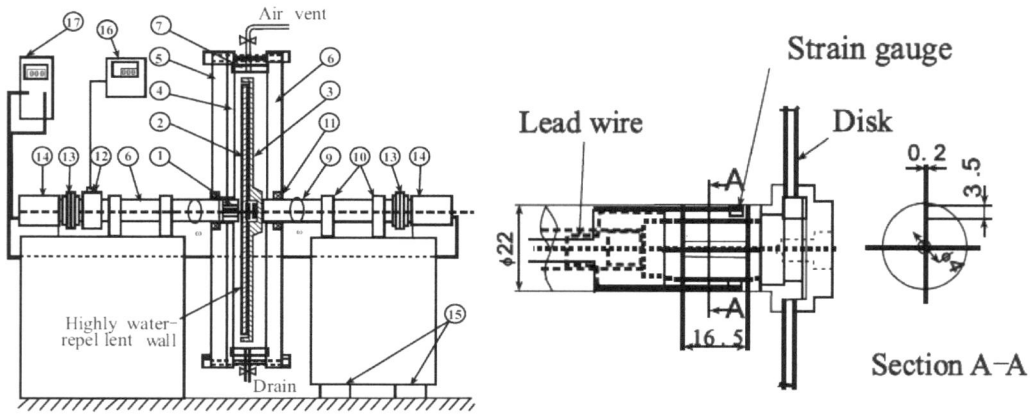

(a) Enclosed rotating disk system   (b) Detailed view of torque measurement device

Figure 9: Experimental apparatus for torque measurement.

The cylindrical housing in which the disk was rotated consisted of spacers ④ housings ⑦ side plate 1 ⑤ and side plate 2 ⑥, which were attached using four bolts. The clearance between the disk ② and the side wall ⑤ was varied to achieve values of 5, 10, and 20 mm in thickness by using three pairs of spacers ④ and housings ⑦. As a result, the experiments were carried out under conditions where the clearance ratios were $s/a = 0.055, 0.111$, and 0.222. The smooth plane disk ② was 180 mm in diameter with a thickness of 3 mm. One side of the walls of the rotating disk was coated with a highly water-repellent material. The coating film thickness and the contact angle for the highly water-repellent wall were approximately $5\ \mu m$ and $120°$, respectively. A smooth aluminum disk that had been used for a no-slip boundary condition was also tested for comparison with the highly water-repellent wall disk. The frictional torque acting on one side of the rotating disk only was measured directly by using the torque measurement device

①, which was a strain gauge cemented to the top of the rotating shaft. As shown in Fig. **9b**, four 0.2-mm thick phosphor bronze plates were soldered to the boss. Four strain gauges were attached to these plates, and the lead wires were inserted into the rotating hollow shaft. The device was water-proofed as a matter of course.

The experimental results for the moment coefficients of an aqueous solution of 30 % glycerin and tap water are shown in Fig. **10a** and **b**, respectively. In these figures, the solid lines are based on the semi-theoretical formulae given by Daily and Nece, which are applied to the case of the disk rotated within a chamber of finite dimensions in the laminar and turbulent flow regions. The dotted lines show Von Kármán's results for laminar flow. These formulae are given in the following equations;

$$\text{Daily \& Nece [3]; } C_m = 1.85 (s/a)^{1/10} \, \text{Re}^{-1/2} \tag{6-9}$$

$$C_m = 0.051 (s/a)^{1/10} \, \text{Re}^{-1/5} \tag{6-10}$$

$$\text{Von Kármán [1]; } C_m = 1.935 \, \text{Re}^{-1/2} \tag{6-11}$$

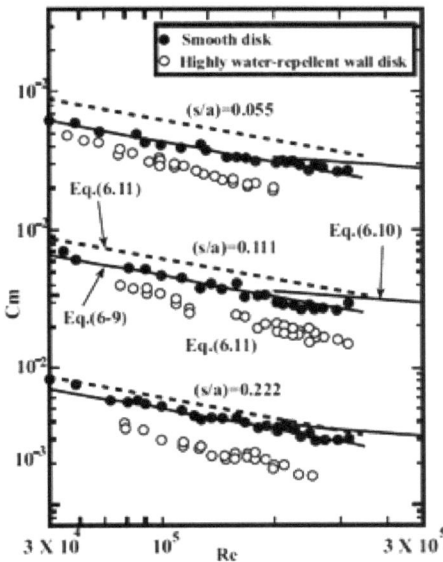

(a) 30wt% Glycerin solution   (b) Tap water

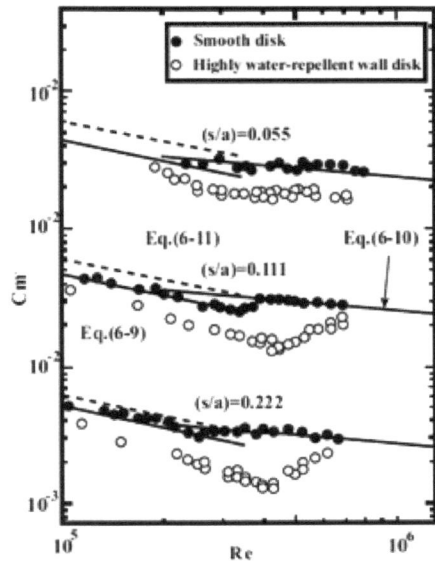

Figure 10: Moment coefficients.

The experimental data for a smooth disk agree well with Eqs. (6-9) and (6-10) for the laminar and turbulent flow ranges, respectively where the no-slip boundary conditions is satisfied. In the case of the highly water-repellent wall disk [10], the moment coefficient was lower than that of the smooth disk where the no-slip boundary condition was not satisfied at the rotating surface. The moment coefficients of the smooth disk are proportional to $Re^{-1/2}$, whereas they are not in the case of the highly water-repellent wall disk. In contrast, the experimental data approach those of the smooth disk as the Reynolds number increases in the turbulent flow range, as shown in Fig. **10b**.

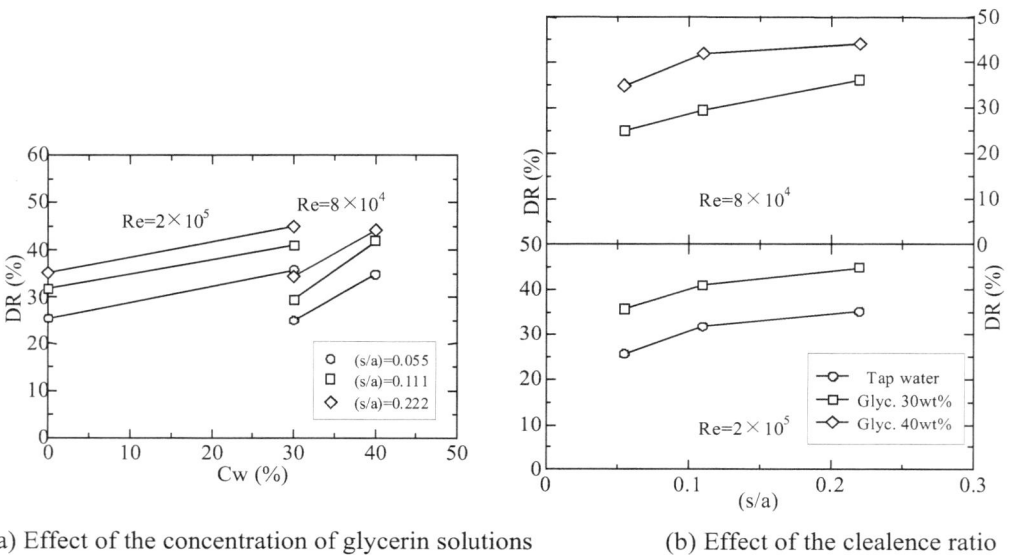

(a) Effect of the concentration of glycerin solutions   (b) Effect of the clealence ratio

Figure 11: Drag reduction ratios of moment coefficients.

This trend is presumably caused by the gas phase separation in the groove on the highly water-repellent wall disk which increases the shear stress acting on the disk surface as described in Chapter 4.

Fig. **11a** and **b** show the effects of the glycerin solution concentration and clearance ratio, respectively on the moment coefficient. In these figures, DR represents the drag reduction ratio based on the moment coefficient of a smooth disk with no-slip boundary conditions. In the case of tap water and a 30wt% glycerin solution at $Re = 2 \times 10^5$, the maximum drag reduction ratios were approximately 35% and 45%, respectively. As shown in Fig. **11b**, the drag reduction tends to increase as a function of the clearance ratio in the experimental range of $(s/a) \le 0.222$. By means of dimensional analysis using these experimental results and Eq. (1-1), we can predict the following functional relation:

$$C_m = f\left(\text{Re}, \frac{s}{a}, \frac{\mu}{s\beta}\right) \qquad (6\text{-}12)$$

where, β is the sliding coefficient given by Eq.(1-1). It is necessary to analyze the moment coefficient using approximate boundary layer theory in order to compare the experimental results quantitatively.

(2) Analytical Approach

In general, the moment coefficient of a rotating disk with the slip boundary condition is analyzed [11] using the momentum integral equation as with the analytical method given by Daily and Nece [3] because the experimental results of the velocity profile in the boundary layer are shown in the previous chapter. Fig. **12** shows the analyzed flow model.

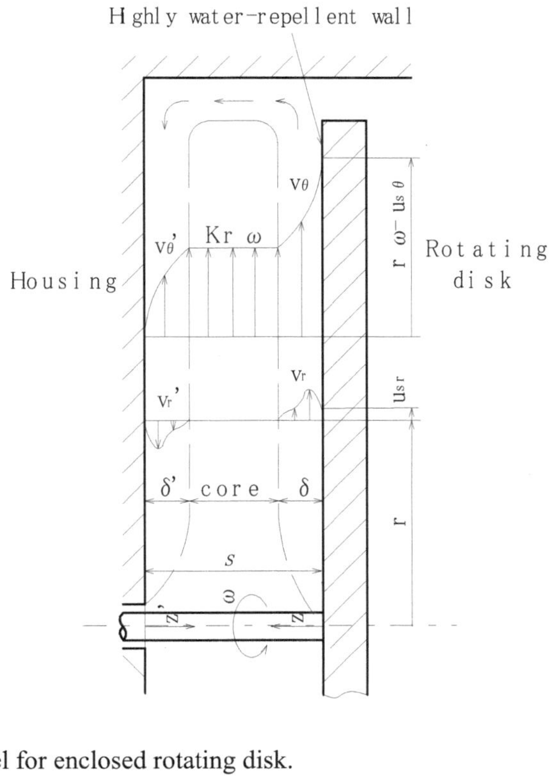

Figure 12: Flow model for enclosed rotating disk.

The moment integral equations [1] of the boundary layer on the rotating disk can be written as

$$\frac{d}{dr}\left(r\int_0^\delta \upsilon_r^2 dz\right) - \int_0^\delta \upsilon_\theta^2 dz = -\frac{\tau_r r}{\rho} \tag{6-13}$$

$$\frac{d}{dr}\left(r^2\int_0^\delta \upsilon_r\upsilon_\theta dz\right) = -\frac{\tau_\theta r^2}{\rho} \tag{6-14}$$

where, υ and τ are the velocity and the shear stress, respectively and the suffix denotes a component of the cylindrical coordinates.

By substituting Eqs. (6-5) and (6-6) into Eqs. (6-13) and (6-14), we can numerically analyze the boundary layer thickness, δ and the term related flow angle in Eq. (6-2), α as the parameters of term related K, the angular velocity ratio between the fluid core and disk, and β, the sliding coefficient in Eq. (1-1).

The wall shear stress are

$$\tau_r = 2\alpha\beta\omega r /(\beta\delta + \mu)$$

$$\tau_\theta = -2\beta\mu\omega r(1-K)/(\beta\delta + 2\mu)$$

In addition, the boundary layer thickness is constant for radius, r; if the effect of disk thickness on the frictional moment of the disk is neglected, the frictional moment M acting on the one side of the disk is given by

$$M = -\int_0^a 2\pi\tau_\theta r^2 dr = \frac{\beta\pi\mu\omega a^4 (1-K)}{\beta\delta + \mu}. \tag{6-15}$$

The moment coefficient is then given by

$$C_m = \frac{M}{\frac{1}{2}\rho a^5\omega^2} = \frac{2\pi\beta(1-K)}{\Delta\beta\sqrt{Re} + \frac{\mu}{a}\cdot Re} \tag{6-16}$$

where,

$$\Delta = \sqrt{\omega\delta^2 / \nu}$$

For a smooth disk with the no-slip boundary condition Eq.(6-16) is also simplified to

$$C_m = \frac{2\pi(1-K)}{\Delta\sqrt{Re}}$$

In addition, the flow angle ϕ is given,

$$\tan\varphi = \frac{\alpha}{1-K}\frac{1+\mu/\beta\delta}{1+2\mu/\beta\delta} \tag{6-17}$$

We notice that α increases with ϕ that is, the flow near the disk trends to the out of the circumference. As might be expected, this agrees qualitatively with the experimental result shown in Fig. **6**.

The results calculated from Eq. (6-16), which show the relationship between C_m and Re as a parameter of β, are shown in Fig. **13**. The value of K = 0.38 is used by referring Fig. **8**. The dotted line shows the semi-theoretical results obtained by Daily and Nice [3], Eq. (6-10).

The experimental data for the 30 wt% and 40wt% glycerin solutions are re-plotted to compare them with these calculated results. A strictly empirical correlation for the sliding coefficient of Eq. (1-1) in Chapter 1 has not yet been sufficiently developed. However, the results give an indication that β ranges from 3 to 5 (Pas/m) within the limits of $1 < \beta < 10$ and the numerical value increases with an increase in the viscosity as shown in Chapte 2 and 4.

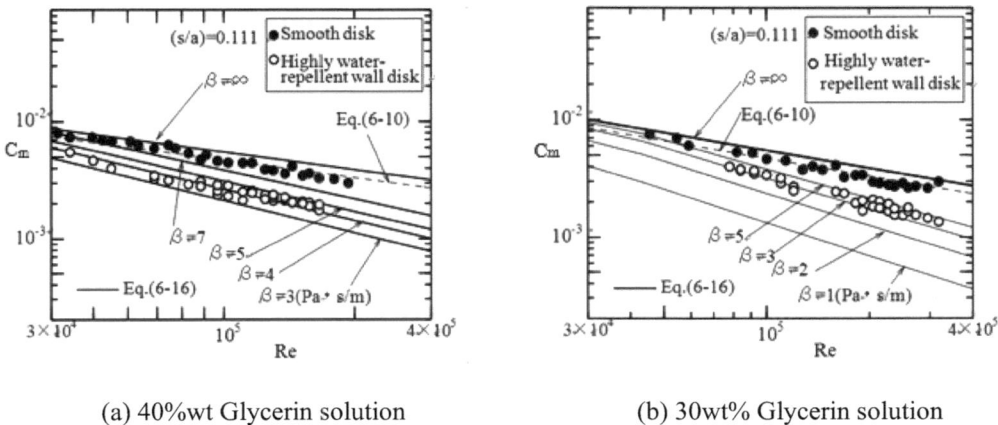

(a) 40%wt Glycerin solution   (b) 30wt% Glycerin solution

Figure 13: Comparison between experimental and analytical results.

SUMMARY

Throughout the forgoing discussion of measurements of the frictional moment acting on an enclosed rotating disk and velocity profile as well as the analytical approach to the moment coefficient by using the integral momentum equation of the boundary layer we have considered the situation in laminar drag reduction where fluid slip occurs at a highly water-repellent or hydrophobic wall from a macroscopic viewpoint. That is, the effect is closely related to a decrease in the tangential velocity near the highly water-repellent wall relative to that in a smooth disk. The flow angle increases near the wall, when the flow pattern changes. Experimental results demonstrate the validity of the effect. However, Estimating the value of the sliding coefficient β applicable at a given condition remains difficult. In principle, this value may be determined by referring to the experimental results.

REFERENCES

[1]   Von Kármán, Th., 1921, "Über laminare und turbulente Reibung", ZAMM, **1**, pp. 233-252.
[2]   Schultz-Grunow, F., 1935, "Der Reibungswiederstand rotierender Scheiben in Gehäusen", ZAMM, **15**. pp.191-200.
[3]   Daily, J. W. and Nece, R. E., 1960, "Chamber Dimension Effects on Induced Flow and Frictional Resistance of Enclosed Rotating Disks", Tran. ASME Series D, 82, pp. 217-232.
[4]   Hoyt, J. W., 1972, "The Effect of Additives on Fluid Friction", J. of Basic Engineering, Tran. ASME Series D, 94, pp. 258-285.
[5]   Watanabe, K., 1978, "Frictional Resistance of a Rotating Disk in Polymer Solutions", Bull. of the JSME, **21**, 153, pp. 455-462
[7]   Ogata, S. and Watanabe, K., 2002, "Limiting maximum drag reduction asymptote for a moment coefficient of a rotating disk in drag reducing surfactant solution", J. Fluid Mech., **457**, pp. 325-337.
[8]   Kato, H., Watanabe, K., and Naya, K., 1978 "Visualization of Fluid Flow Near a Rotating Disk in Dilute polymer Solutions", Bull. of the JSME, **21**, 161,pp.1618-1625.
[9]   Cochran, W.G., Proc.1934, "The flow due to rotating disk", Proc. Camber. Phil. Soc., 30, pp.365-375.
[10]   Watanabe, K. and Ogata, S., 1997, "Drag Reduction for a Rotating Disk with Slip in Newtonian Fluids,"Proc. of JSME International Conference on Fluid Engineering", pp. 545-550.
[11]   Watanabe, K., and Ogata, S., 1998, "Drag Reduction for a Rotating Disk with Highly Water repellent Wall", JSME Int. J. Ser. B, 41, 3, -
[12]   Ogata, S and Watanabe, K., 1999, "Flow Characteristics of a Drag reducing Rotating Disk with Highly Water Repellent Wall", Proc. of Sym. on Flow Control of Wall-Boundary and Free Shear Flows, Joint ASME/JSME Fluids Eng. Conf., FEDSM99-6936, pp.1-6.

Laminar Drag Reduction, 2015, 71-82

CHAPTER 7

Flow Past a Circular Cylinder

Abstract: Flow patterns for a circular cylinder with fluid slip have been described experimentally using a cylinder with a highly water-repellent surface with Reynolds numbers Re-ranging from Re = 20 to 150. In addition, the stream lines of the flow past a circular cylinder have been analyzed in the Reynolds numbers 20 and 50 by using the fluid slip boundary condition described in Chapter 1. The analytical results agreed well with the experimental results obtained by the flow visualization. Flow separation can often result in increased increasing drag, particularly pressure drag which is caused by the pressure differential between the front and rear surface of the cylinder as it travels through the fluid. The delay in the flow separation is associated with a significant reduction in the drag. The results showed that the separation point of a cylinder with a highly water-repellent surface moves downstream compared to that of a smooth surface cylinder. This phenomenon is conducive to drag reduction with a fluid slip. The drag reduction ratios for tap water obtained from this calculation are 15% and 10% at Re = 20 and 50, respectively. The Strouhal number of a cylinder with fluid slip increases in comparison with the value for a smooth surface cylinder with no fluid slip.

Keywords: Drag reduction, external flow, circular cylinder, flow drag, highly water-repellent wall, Strouhal number, wake, separation, fluid slip, drag coefficient, velocity defect, streamline.

1. INTRODUCTION

Flow past a circular cylinder is one of the most simple and important flows around a bluff body. This type of flow happens practically in one's own backyard such as the flow around an overhead contact electric wire, a bridge column, a circular structural object or a heat exchanger tube, *etc*. Thus, analysis of this flow is highly relevant to achieve breakthroughs in real-world engineering application problems and it will be necessary to estimate the total drag acting on a circular cylinder to assure the security and reliability of these objects in structural designs. The total drag of a circular cylinder in a real fluid is given by the sum of the viscous and the pressure drags due to the pressure difference between the front and rear surface of the cylinder as it travels through the fluid. For an infinite circular cylinder, the total drag depends on the Reynolds number only. For a low Reynolds number, *e.g.*, less than 3×10^5, viscous forces dominate and the boundary layer on the surface may generally be considered laminar. The boundary layers and wake play a major role in the pressure distribution on the cylinder; therefore, it is very important to understand the flow pattern in order to calculate the drag. The flow pattern becomes two- dimensional when it is infinite in the length and is oriented vertically with respect to a steady uniform flow, and this simple geometry demonstrates benefit from the experimental observation. Many previous studies on the flow past a cylinder have been done hitherto and it is well known that the variation of the drag coefficient [1-5] with the Reynolds number for Newtonian fluids can be described by a single curve.

The previously reported these results for circular cylinder in Newtonian fluids have been obtained by applying the no-slip boundary condition on the cylinder surface. However, there are very few studies which clarified experimentally the flow pattern relating fluid slip at a cylinder surface, and the change in the pattern has a certain impact on the drag of a cylinder. Flow around a circular cylinder has many flow patterns because the boundary layer is laminar over large Reynolds number range. Thus, it is a useful flow field to research the effect of fluid slip on laminar drag reduction phenomena. In addition, the flow past a circular cylinder is convenient for experimentally examining the effect of fluid slip on the flow behavior at near the surface by the experiment because we can easily visualize the flow using a water tunnel. If it is at all possible to decrease the hydrodynamic force by the drag reduction, it will be conducive the durability improvement or the weight saving of these objects.

In this chapter, we report the experimental flow visualization and velocity measurement results [6] obtained by the flow visualization and the velocity measurement for a circular cylinder with a highly water-repellent surface in the Reynolds number range of 20~150; some interesting results relating to vortex formation in this Reynolds number range are presented. The flow patterns around a circular cylinder with fluid slip were obtained by solving the equations of the stream-function and the vortices. The analytical results for a flow around a circular cylinder with fluid slip suggests a qualitative explanation for the reasons for the laminar drag reduction in laminar drag coefficient for a circular cylinder with a highly water repellent surface [6].

2. EXPERIMENTAL APPROACH

2.1. Flow Visualization

Experiments were conducted in a water tunnel which has a test section with width, length and depth dimensions of 400 mm $\times$ 450 mm $\times$ 300, respectively. The tunnel is of the closed return type and has a free stream turbulence level of about less than approximately 1%. It was approximately 10 mm in the extent of the impact on the uniform velocity profile for the effective boundary layer on the bottom. Fig. **1** shows the test section of a cylinder in the water tunnel. The cylinders used in the investigation were 20 mm in diameter. They were made of bronze and two circular cylinders with a smooth surface and one with a highly water-repellent surface were examined to visually compare the flow around them. The flow patterns were visualized using a dye streak and the hydrogen bubble method for Re = 20, 50, 100 and 150, and photographs of the flow patterns were taken with a digital camera. Dyestuffs were used as the tracer and sedimentation did not occur during the measurement. In the hydrogen bubble method, the negative pole wire was a 50μm diameter platinum wire. The effect of the floating velocity of the bubble on the measured value can be neglected because the order for the main stream velocity is 2 ~ 4 % if we calculate it by using the Stokes' law. The circular cylinder with a highly water-repellent wall is a coating cylinder with a poly-tetra-fluoride-ethylene coating. This is the same coating material used in the case of the circular pipe obtained by laminar drag reduction obtained. Test fluid is tap water with a constant temperature of 14 ℃. The flows examined in this study, are in the flow patterns of two attached eddies behind the cylinder at Re = 20 and 50, and the flow patterns at Re = 100 and 150, known as Kármán vortex street. The velocity profiles of the wake were measured at Re = 20 and 50 using a hot film anemometer at an interval of 1 mm to the distance of the width of the test section. The Strouhal number was measured using the probe set at the position of six times of the cylinder radius in the Re = 100 and 150 cases. Experimental results for the flow pattern are shown in Figs. **2a** and **b**. The wake length [7, 8] increases with increasing Reynolds number, and it is known that a twin vortex occurs in the wake in the range of 8 ≦ Re ≦ 50. The wake of the circular cylinder with the highly water-repellent surface increases in length compared to that of the smooth surface circular cylinder, as shown in Fig. **2a**.

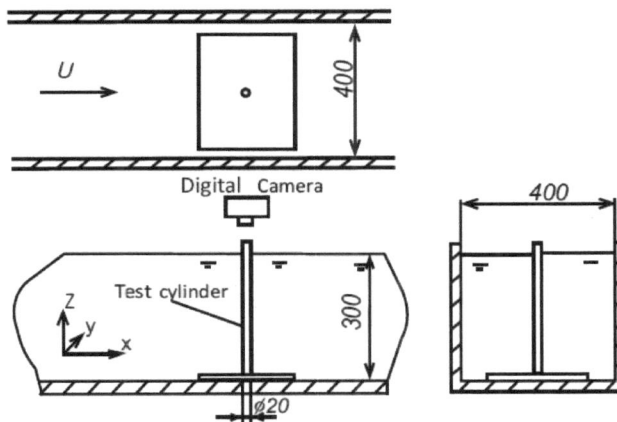

Figrue 1: Test section of water tunnel.

(a) Re = 50

(i) Smooth surface

(b) Re = 150

(i) Smooth surface

(ii) Higly water-repellent surface

(ii) Higly water-repellent surface

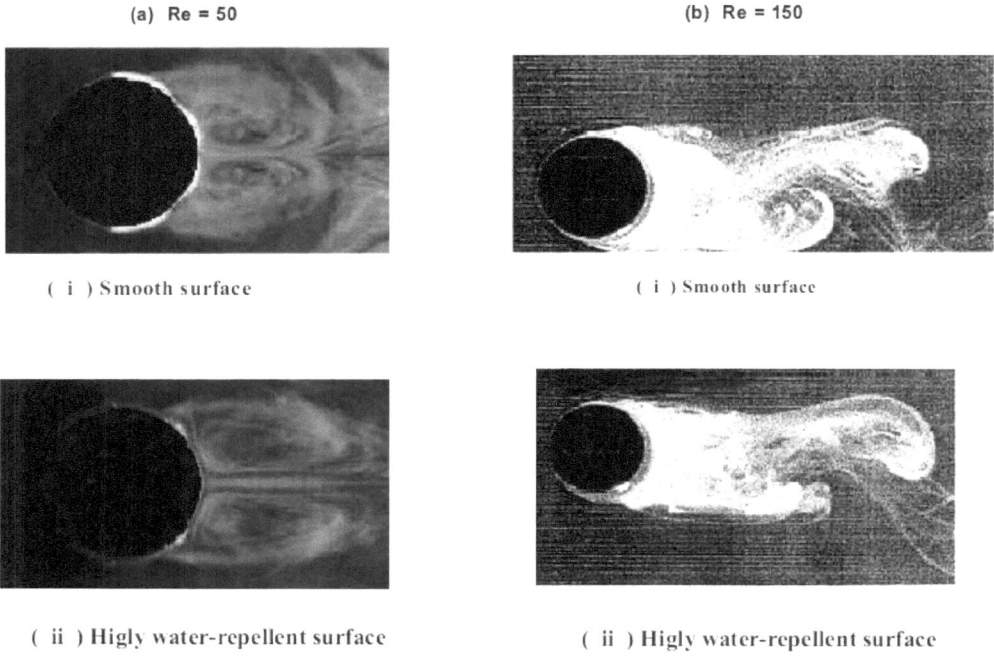

Figure 2: Flow around a circular cylinder.

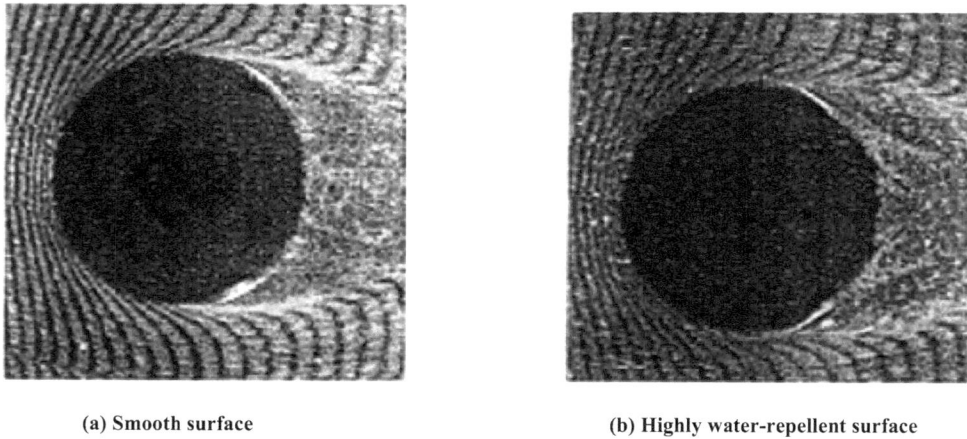

(a) Smooth surface

(b) Highly water-repellent surface

Figure 3: Flow visualized by a hydrogen bubble method at Re = 50.

As the results of the increase in the Reynolds number, a Kármán vortex street occurs in the wake as shown in Fig. **2b**. The effect of fluid slip along the highly water-repellent surface cylinder is that the position of the Kármán vortex street moves to the downstream direction. In both the cases, the observed phenomena indicate a decrease in the velocity defect in the wake. General by speaking, the fluid in the boundary layer is accelerated because the pressure decreases along the cylinder wall from the stagnation point in the downstream direction, and the fluid slows down behind the cylinder because the pressure increases along the downstream direction. The original small kinematic energy fluid in the boundary layer does not flow to downstream smoothly, and the separation occurs at the surface on the back side of the flow as shown in Fig. **3**. The separation angle almost agreed with the value for Newtonian fluids [4] under a no-slip boundary condition. However, the separation point of the highly water repellent surface cylinder moves in the downstream direction as a result of the fluid slip. Separation angle increases by approximately $5\sim10°$ in comparison to that of the smooth surface cylinder. The separation data obtained in this investigation are expected to form the basis for related numerical studies as well as comparison for their accuracy and validity.

2.2. Velocity Profile of Wake

Measurements of the velocity profile of the wake are performed at 1 mm intervals in the y direction using a hot film velocity meter for Re = 20 and 50. The back flow is measured by inverting the probe in the zero velocity regions. Fig. **4** shows the velocity profiles of the wake at the positions from the center of the cylinder to $x = 2r$, 20 mm. The region of decreasing velocity is the velocity defect zone, in each Reynolds number case, this zone has a narrow width for the circular cylinder with the highly water repellent-surface in comparison to that of a smooth surface cylinder. This means that the drag acting on the circular cylinder with highly water-repellent surface comparing with that of a smooth surface cylinder.

We can calculate the drag using the measured value of the velocity defect in the wake by applying the momentum balance. Thus, the drag D per unit length acting on a two dimensional circular cylinder is given as follows:

$$D = \rho \int_{-\infty}^{+\infty} U_m (U_m - U) dy$$

$$(7\text{-}1)$$

where, ρ is the fluid density. U_m and U are the main uniform and the wake velocities, respectively.

By the graphical integration of Fig. **4**, the drag is calculated using Eq. (7-1). The moment coefficient of a cylinder C_d is plotted with the lozenge symbols in Fig. **5**.

The solid line denotes the previously reported results for the drag coefficient of a smooth surface circular cylinder with no fluid slip. The data for the circular cylinder with the highly water-repellent surface show a decrease of 25%~15% as compared to that of a smooth surface circular cylinder. Consequently, the delay in the separation flow that occurs as a result of fluid slip is conducive to the drag reduction in each Reynolds number case.

It is important to clarify the characteristics of vortex shedding in the wake in order to understand the laminar drag reduction in a flow past a circular cylinder.

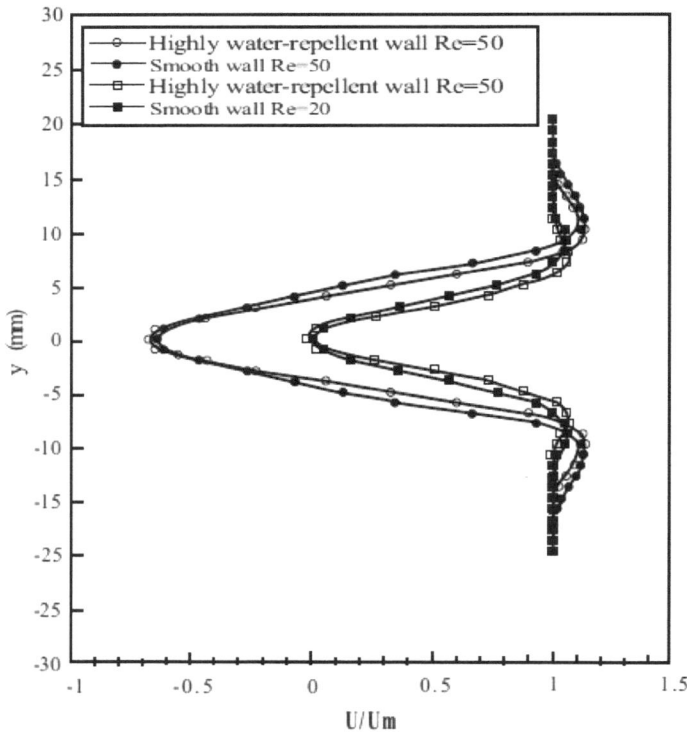

Figure 4: Velocity profiles for the wake of circular cylinder.

Fig. **6** shows the relationship between the wake length and the Reynolds number. Twin vortices are simultaneously shed from both sides of the cylinder in this Re range. The wake field was examined using a hot wire probe. Although the experimental data for a smooth cylinder obtained in this study nearly agree with that of other studies [7, 8] at Re = 20, it is less than the data reported in those studies for Re = 50. In the case of a circular cylinder with a highly water-repellent surface, the length is greater than that for a smooth circular cylinder as shown in Fig. **6**. The impact of the fluid slip is that the wake is elongated. This phenomenon supports the idea that a decrease in the velocity defect increases the velocity as shown in Fig. **4**.

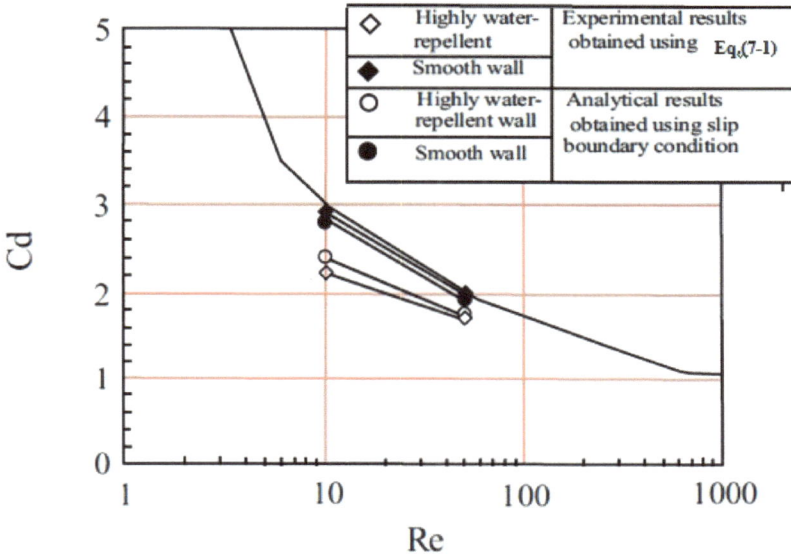

Figure 5: Drag coefficients of a circular cylinder.

As a function of the increase in the Reynolds number, the flows have an oscillatory pattern which depends on Re. The phenomenon of periodic vortex shedding behind a circular cylinder, referred to as a Kármán vortex street is commonly understandable using the Strouhal number, determined by dimensional analysis. The Strouhal number is given as:

$$St = \frac{fd}{U_m} \tag{7-2}$$

where f, d and U_m are the shedding frequency, the diameter of the circular cylinder and the uniform velocity, respectively. Fig. **7** shows the experimentally determined Strouhal numbers. The solid line in the figure denotes Roshiko's experimental results [4] for a smooth surface circular cylinder. It is seen that the Strouhal number for a circular cylinder with a highly water-repellent surface increases.

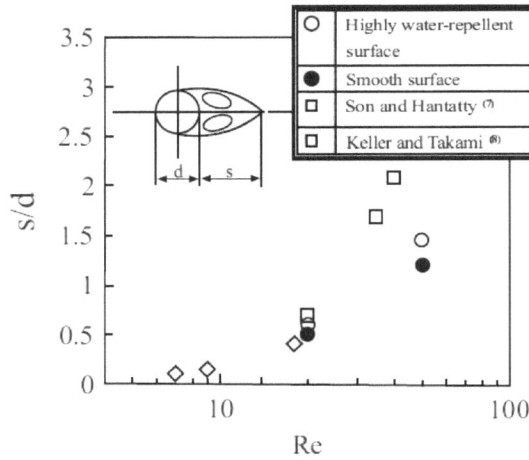

Figure 6: Wake lengths of a circular cylinder.

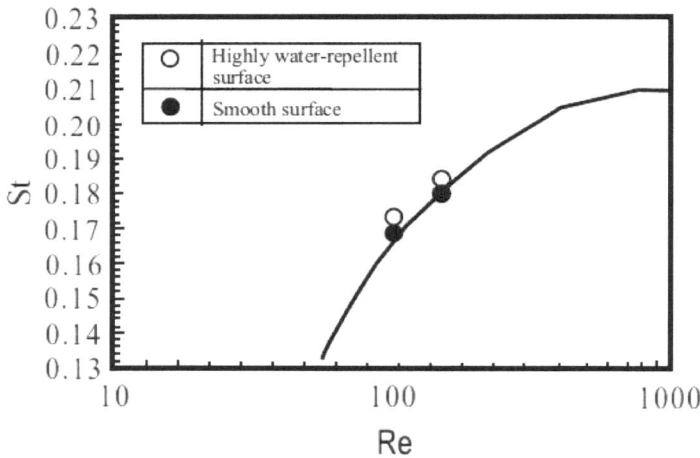

Figure 7: Strouhal numbers.

3. ANALYTICAL APPROACH

In general, the drag coefficients acting on a cylinder at the Re = 20 and 50 are calculated in two ways: (1) an analytical determination using the pressure profile obtained from the numerical calculation, and (2) from the experimental results of the velocity profile for the wake using the momentum equation. The latter method is described above in the Section 2-2, and the results are plotted in Fig. **5**. The flow visualization results showed that the twin vortex of the cylinder with the highly water-repellent surface is increased in length compared to that of a smooth cylinder.

In this section, we discuss about the flow pattern [6] by comparing the experimental visualization result with the analytical result of the stream line with a fluid slip.

We consider a flow in which a cylinder with radius 1 and infinite length is set in an infinite region and the fluids uniformly flow perpendicularly to the center axis of the cylinder. For an incompressible fluid, the equation for the stream function, φ, and the vorticity, ω, which is written in polar coordinates are given as follows:

$$\frac{\partial^2 \varphi}{\partial \xi^2} + \frac{\partial^2 \varphi}{\partial \theta^2} = -\omega e^{2\varsigma} \tag{7-3}$$

$$\frac{\partial \omega}{\partial t} + e^{-2\xi}\left(\frac{\partial \varphi}{\partial \xi}\frac{\partial \omega}{\partial \theta} - \frac{\partial \varphi}{\partial \theta}\frac{\partial \omega}{\partial \xi}\right) = \frac{e^{-2\xi}}{Re}\left(\frac{\partial^2 \omega}{\partial \xi^2} + \frac{\partial^2 \omega}{\partial \theta^2}\right) \tag{7-4}$$

The boundary conditions of the stream function at the cylinder wall and a free-stream region are:

$$r = a; \quad \phi_0 = 0 \tag{7-5}$$

$$r \to \infty; \quad \phi_\infty = y = e^\varsigma \sin\theta \tag{7-6}$$

For the vorticity, because the flow is a uniform stream at an infinite distance it becomes:

$$r = a; \quad \omega_0 = -e^{-2\xi}\partial^2\psi/\partial\xi^2 \tag{7-7}$$

$$r \to \infty; \quad \omega_\infty = 0 \tag{7-8}$$

From the finite-difference approximation of Eq. (7-6), we obtain the following equation by applying $\xi=0$ and $\varphi=0$ at the cylinder wall:

$$\omega_o = -(\phi_{k+1} + \phi_{k-1})/(\Delta\xi)^2 \tag{7-9}$$

where φ_{k+1} and φ_{k-1} are the grid points. The slip velocity at the cylinder wall is given as:

$$\upsilon_\theta = (\phi_{k+1} - \phi_{k-1})/\Delta\xi = u_s, \tag{7-10}$$

where, u_s is the slip velocity:

By combining Eqs. (7-9) and (7-10), we get:

$$\omega_o = (u_s \Delta \xi - 2 \phi_{k+1})/(\Delta \xi)^2 . \qquad (7-11)$$

The initial conditions are:

$$\phi = (r - (\frac{1}{r})) \sin \theta = (e^{\xi} - e^{-\xi}) \sin \theta , \qquad (7-12)$$
$$\omega = 0$$

Using of these boundary conditions, Eqs. (7-3) and (7-4) are solved numerically using a finite-difference model. For the model we used a polar coordinate mesh where the mesh dimensions were 60×40. The slip velocity must be given at the wall in the calculation. The slip velocity is calculated by Eq. (1-1) described in Chapter 1. Thus, the slip velocity, u_s is given by the following equation:

$$\tau_w = \mu \left| \frac{\partial u}{\partial y} \right|_{y=0} = \beta u_s \qquad (7-13)$$

where μ, u_s, and β are the viscosity, the slip velocity at the wall and the sliding coefficient related to the viscosity, μ and the physical property of the wall surface, respectively. The equation reduces to the no-slip boundary condition if we let β equal infinity. After assuming the value of the sliding coefficient β, the slip velocity was introduced into the numerical calculation step using the above equation.

Fig. **8a-d** show the streamlines of the numerical solutions for the flow past a circular cylinder with Re = 50. The flow in Fig. **8a** in the case of the no-slip condition shows that the standing eddies, whose size increases with Reynolds number, have formed behind the circular cylinder. This flow pattern agrees well with the experimental visualization result of Fig. **2a**.

In the case of the fluid slip, the results indicate that the standing eddies are elongated to the downstream direction for less $\beta = 4$ (Pa · s/m). As a function of the increase of β, the eddy size becomes small, as shown in Fig. **8d**. This result means that the flow pattern resembles the flow of an ideal fluid with an increase of the slip velocity. As described in Chapter 1, it is difficult to make a prediction for the value of the sliding constant β in Eq. (7-13) because the experimental data for the fluid slip are lacking at present. However, we can calculate the order of β from some analytical results on the slip flow by comparison to experimental results. Thus, the result produces meaningful results for the calculation of Eq. (7-13). The value is the range in value which was assumed for the flows in the pipe in Chapter 2 and the enclosed rotating disk in Chapter 6. Thus, the analytical result for the drag coefficient of the cylinder with fluid slip for the case of $\beta = 4$ (Pa · s/m), is plotted with the white circle symbol in Fig. **5**. These data show a reduction of approximately 15% and 10% in the drag reduction ratio at Re = 10 and 50, respectively. In the wakes behind a cylinder, the rear twin-vortices begin to form at Re=5, then become more enlarged as the Reynolds number increases, and finally become asymmetrical at approximately Re=50, as shown in Fig. **8**. On the other hand, as a function of an increasing Re, over a large range of Reynolds number 50-10^5, eddies are formed continuously from each side of a cylinder which is well known as the Kármán vortex street. It is interesting to examine the change in the eddy which occurs with fluid slip. This problem will be described in Chapter 8.

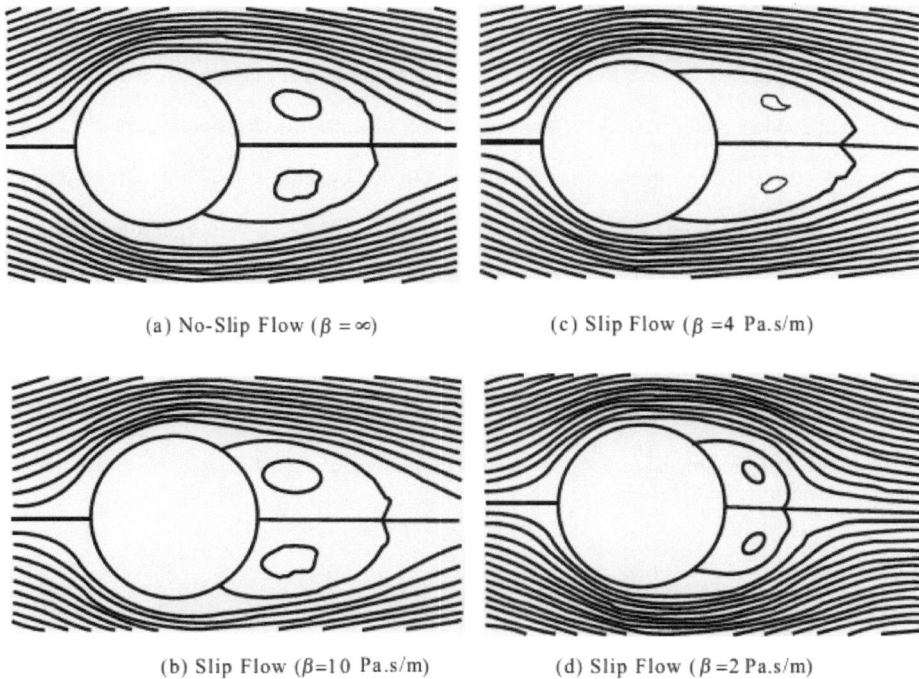

(a) No-Slip Flow ($\beta = \infty$)   (c) Slip Flow ($\beta = 4$ Pa.s/m)

(b) Slip Flow ($\beta = 10$ Pa.s/m)   (d) Slip Flow ($\beta = 2$ Pa.s/m)

Figure 8: Calculated results of stream lines of a circular cylinder at Re = 50.

SUMMARY

Flow past a circular cylinder generates a separation, a wake, and vortex shedding with an increase in the Reynolds number. The effect of fluid slip at the surface was experimentally and theoretically investigated for the flow patterns of a circular cylinder with a highly water-repellent surface. From the velocity measurement of the wake behind a circular cylinder with a highly water-repellent wall, it was observed that a drag reduction of the drag coefficient of approximately 25% and 15% occurs at Re=10 and 50, respectively. These conclusions imply that the separation point in the flow moves downstream by fluid slip on the cylinder surface, and these phenomena were examined using a flow visualization of the flow pattern around the circular cylinder. For the calculated stream lines by applying the boundary condition of the slip velocity at the surface, the calculated flow patterns agree well with ones observed in the the experimental photo-traces. The calculated drag coefficients for Re = 10 and 50 are reduced by approximately 15% and 10% if we assume a value for the sliding coefficient of $\beta = 4$ pa·s/m.

REFERENCES

[1]   Wieselsberger, C., 1921, "Neuere Feststelleungen über die Gesetze des Flüssigkeits- und Luftwiderstands" Phys. Z. 22, pp. 321-328.

[2]   Finn, R. K., 1953, "Determination of the drag on a cylinder at low Reynolds numbers", J. Applied Physics, **24**, no. 6, pp.771-773.

[3]   Tritton, D.J., 1959, "Experiments on the flow past a circular cylinder at low Reynolds numbers", J. of Fluid Mech., **6**, pp. 547-567.

[4]   Roshko, A. 1961, "Experiments on the flow past a circular cylinder at very high Reynolds number"., J. of Fluid Mech., **10**, pp. 345-356.

[5]   Pruppacher, H.R., Clair, B. P. L. and Hamielec, A. E. 1970, " Some relations between drag and flow pattern of viscous flow past a sphere and a cylinder at low and intermediate Reynolds numbers ", J. Fluid Mech., **44**, part 4, pp.781-790.

[6]   Watanabe, K. and Fujita, T., 2000, "Drag Reduction of a Circular Cylinder with a Highly Water-Repellent Wall", Trans. of the JSME, Ser. B. 66-650, pp.53-58.

[7]   Son, J. S. and Hanratty, T. J. 1969, "Numerical solution for the flow around a cylinder at Reynolds numbers of 40, 200 and 500". J. Fluid Mech., **35**, part 2, pp.369-386.

[8]   Keller, H. B. and Takami, H., 1966, "Numerical Solutions of Non-Linear Differential Equations", Proc. of Adv. Symp. pp.115-121.

Laminar Drag Reduction, 2015, 83-103

CHAPTER 8

Flow Past a Sphere

Abstract: Drag reduction for a uniform velocity flow past a sphere with highly water-repellent surface and a gas liquid interface has been investigated experimentally and analytically for Reynolds numbers ranging from 10 to 10^4. The surface of a sphere had high water repellency (contact angle of ~150°) and the gas liquid interface exists at the surface with many fine grooves. Experimental results showed that the separation point of the boundary layer around a sphere moved downstream compared with that of a smooth surface sphere. It was also shown, by measuring the sphere drag, that drag reduction for a sphere with highly water-repellent surface occurs at Reynolds numbers less than 10^4 and that the maximal drag reduction ratio is 28.5 % at $Re = 7.2$. By considering that such a phenomenon occurs due to an apparent fluid slip at the gas liquid interface of a hydrophobic surface, the flow patterns around a sphere were analyzed by applying the gas liquid two-phase model at the surface proximity. Results of numerical simulations were obtained for Reynolds number ranging from 100 to 450. The boundary condition for fluid slip was given by assuming an effective slip boundary condition of the surface. A comparison of the simulation results with the experimental results shows a close agreement concerning the flow patterns of the wake and drag coefficient.

Keywords: Drag reduction, external flow, sphere, flow drag, highly water-repellent wall, wake, gas-liquid interface, numerical simulation, fluid slip, drag coefficient, falling ball test, flow pattern, separation, pressure profile.

1. INTRODUCTION

A sphere is frequently used as the simplest model of a particle in the multiphase flow, and we can apply this model to measure the liquid viscosity by using Stokes' law, which, as is well known, is derived from Navier-Stokes equation for very small Reynolds numbers. Because a sphere is a simple symmetrical configuration of a bluff body, it is convenient to grasping the flow-surface interaction by flow visualization, in which the separation or wake occurs. Thus, from many previous studies, the relationship between the drag coefficient and Reynolds number is quantitatively given for a wide range of flow types. However, these results are obtained using no fluid slip for the boundary condition on the surface, and currently, few studies on the flow characteristics of a sphere with fluid slip exist.

As was shown in the preceding chapters, (Chapter 1-7), the analytical results on the friction loss or drag of laminar flow with fluid slip are relatively easy to obtained from Navier-Stokes equation using Navier's hypothesis (see Chapter 1) and as shown before the analytical results are in a good qualitative agreement with the experimental results. However, it is difficult to estimate the sliding constant in Navier's hypothesis. Although the governing equation is very simple, the idea of using the external friction analogy to solid friction for fluid slip may have its limitations. As was described in the preceding chapters, the apparent slip velocity is observed in some internal or external flows on highly water-repellent surface that have many fine grooves. Visualization of flow in the proximity of a surface disclose that the existence of gas phase in fine grooves is a very important factor that contributes to the development of a fluid slip. Therefore, for proper modeling and numerical simulations, it is necessary to account for the existence of gas liquid-solid interface. In this chapter, the drag reduction phenomena [1] for a sphere drag with fluid slip is obtained experimentally, and the flow past a sphere with highly water-repellent surface with many fine grooves is analyzed by using the gas liquid two-phase flow model [2]. Results of this analysis are compared with the experimental results that are obtained by the drag measurements and flow visualization [3].

2. EXPERIMENTAL APPROACH

2.1. Flow Visualization

Flow visualization experiment was performed in a vertical recirculation water tunnel equipped with a plexi-glass test section that was 300 mm wide, 320 mm high and 1,190 mm long as shown in Fig. **1**. Mean velocity in the test section ranged from 0.0033 m/s to 0.139 m/s. Velocity in the section had a uniform profile, with a scattering of ±1.51%. The test sphere was attached to a support rod that was set up 150 mm downstream of the test section' front. The flow was visualized by injecting a water-based dye upstream of the nozzle. Images of the flow patterns obtained following the dye injection were recorded using a digital video camera at a recording rate 30 frame/s. Sphere was tested for diameter ranging from 2.38 mm to 8.57 mm. The values of Reynolds number ranged from 10 to 10^4. A sphere with a highly water-repellent surface was made by spray coating (HIREC 450, NTT-AT Co.,) the surface with hydrophobic paint. The contact angle of the surface was about 150°. Surface roughness of the coating is shown in Fig. **2**. The ruggedness of the surface was ±15μm.

When the test sphere was placed on the surface of the water in a vessel a very interesting phenomenon was observed, as shown in Fig. **3**. The behaviors of the

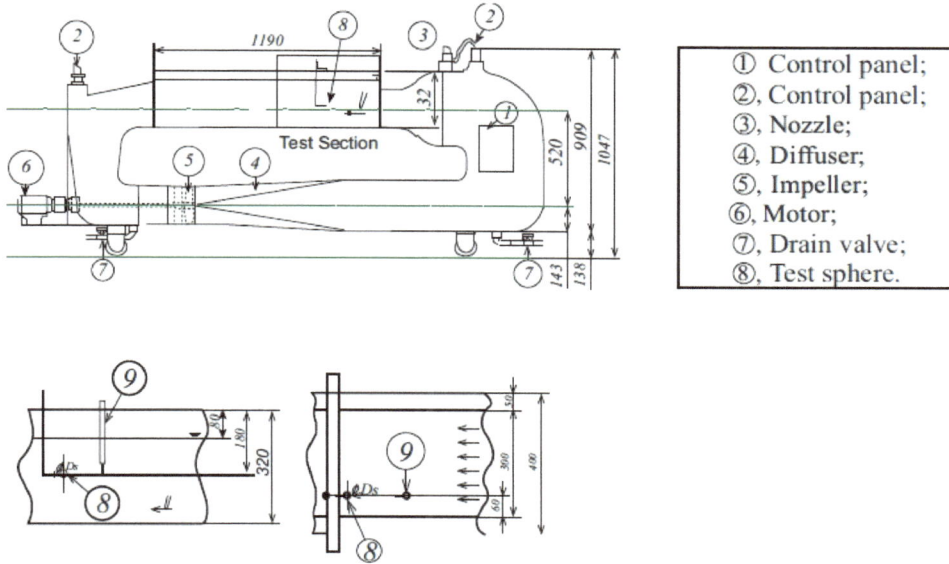

Figure 1: Water tunnel.

sphere in stationary water could be characterized by two patterns. In one of the patterns, shown in Fig. **4a**, the sphere did not settle even if its density became higher than that of the water, with increasing interfacial tension of the hydrophobic surface. In the second pattern, shown in Fig. **4b** the sphere floated or settles down because it was wrapped by the gas phase film when it was in the water. The condition in which the sphere has rest or is falling is dependent on the spheres and density difference of the liquid, the interfacial tension, the thickness of the gas phase and the sphere diameters. The equilibrium equation of forces acting on the sphere is analyzed and the discriminant diagram [4] is obtained.

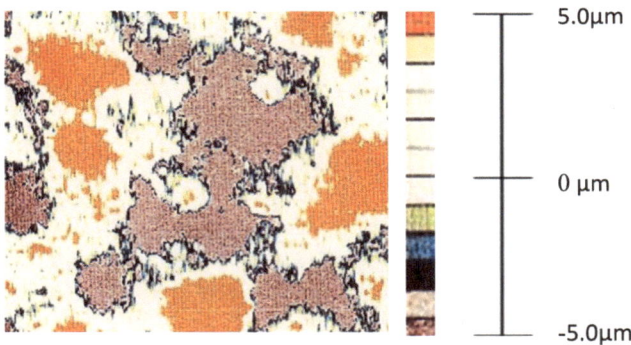

Figure 2: Surface roughness.

We measured the interfacial tension and the thickness of the gas phase film for the first and second scenarios, respectively. The thickness of the gas phase film [4] was almost constant and depended on the material from which the sphere was made but not on the sphere's diameter. The gas phase thickness values were 61 μ m and 91 μ m [1] for polyacetal (commercial name, Duracon) and nylon spheres, respectively.

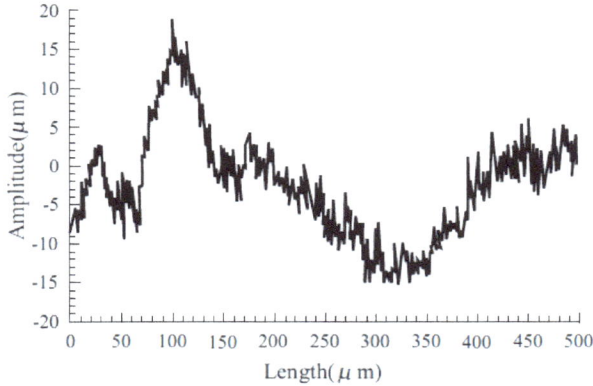

Figure 3: Quantification of the surface roughness.

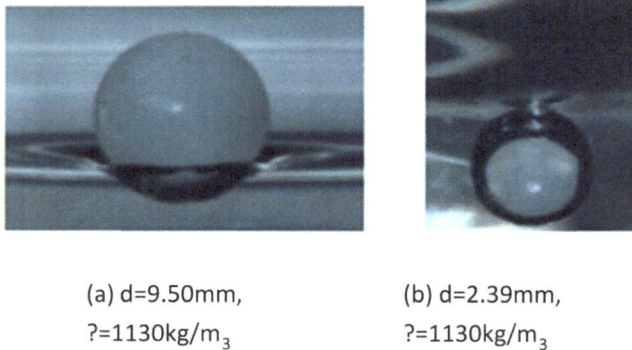

| (a) d=9.50mm, | (b) d=2.39mm, |
| ?=1130kg/m$_3$ | ?=1130kg/m$_3$ |

Figure 4: Sphere with highly water-repellent surface in rest water.

2.2. Drag Measurement

Fig. **5** shows the experimental apparatus used for the drag measurements. An experiment on the drag of a sphere was carried out using an acrylic tank filled with tap water. The aim was to measure the fall velocity of a free-falling sphere starting from rest. The height of the acrylic tank was 1,500 mm. The fall velocity of the test sphere was calculated from images captured with a digital video camera. The fall velocity of a sphere was calculated for a smooth surface sphere and for a sphere with highly water-repellent surface by using the momentum equation. Considering the thin gas phase film, this equation is written

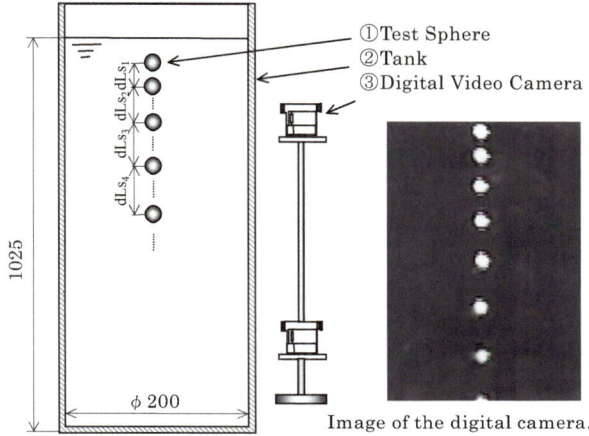

Figure 5: Falling sphere test setup.

$$\{(m_s + m_{gas}) + K m_f\}\frac{dV_s}{dt} = \{(m_s + m_{gas}) - m_f\}g - C_d \pi \rho_f \frac{(D_s + 2T_g)^2}{8} V_s^2$$

$$- \frac{3}{2}(D_s + 2T_g)^2 \sqrt{\pi \rho_f \mu} \int_0^t \frac{(dV_s / dt)}{\sqrt{t - \tau}} d\tau$$

$$(8\text{-}1)$$

where, m, V_s and D denote the mass, fall velocity and diameter, respectively. The subscript s and f denote the sphere and the liquid, respectively. The mass and the thickness of the gas phase that surrounding the sphere are denoted by m_{gas} and T_g, respectively. μ is and viscosity of the fluids. The drag coefficient Cd in Eq. (8-1) is given by following equations;

Stokes [5]

$$C_d = \frac{24}{Re}$$

$$(8\text{-}2)$$

$$Re \langle 1$$

Schiller & Naumann [6]

$$C_d = \frac{24}{Re}\left(1 + 0.15\,Re^{0.687}\right)$$

(8-3)

$$1 \le Re \langle 700$$

Carey [7]

$$C_d = \left[\left(\frac{24}{Re}\right)^{1/2} + 0.03435\left(Re^{0.06071} + \frac{1}{1.72013 + 0.018\,Re}\right)\right]^2$$

(8-4)

$$700 \le Re \langle 10443$$

Brauer [8]

$$C_d = \frac{24}{Re} + \frac{3.73}{Re^{1/2}} - \frac{4.83 \times 10^{-3}\,Re^{1/2}}{1 - 3 \times 10^{-6}\,Re^{3/2}} + 0.49$$

(8-5)

$$10443 \le Re \langle 100000$$

Fig. **6** shows the values obtained using these equations for a range of Reynolds numbers, and it is known that these values agree with the experimental result obtained for normal smooth sphere with no slip, over the entire Reynolds number range.

In this experiment, polyacetal and nylon spheres were tested, and the densities of these spheres were $\rho_s = 1362$ kg/m^3 and 1131 kg/m^3, respectively. The diameters of the test spheres ranged from 2.38 mm to 19.00 mm, and the Reynolds numbers ranged from 5 to 10^4.

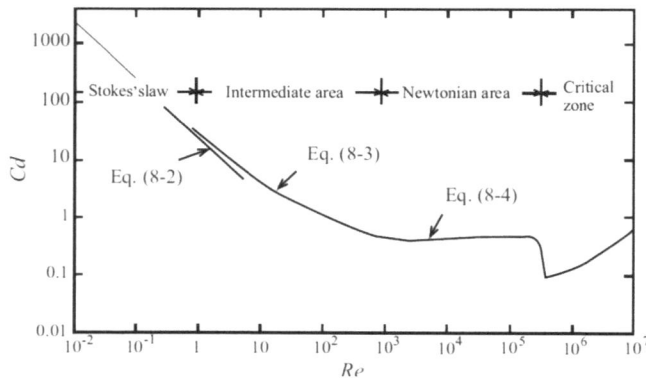

Figure 6: Drag coefficient of a sphere.

3. EXPERIMENTAL RESULTS

Wakes generated by the sphere were investigated at Reynolds numbers ranging from 10 to 430. Fig. **7** shows typical photographs of the vortex ring behind a sphere. For sphere with a normal smooth surface, a permanent vortex ring begins to form behind the sphere for Reynolds numbers greater than or equal to about 30, and the end of the wake begins to oscillate when Re is about 130. In contrast, for spheres with the highly water-repellent surface, a permanent vortex ring begins to form behind the sphere at a Re of about 30, and the end of the wake begins to oscillate when Re is about 100. Moreover, the vortex ring becomes narrow. It can be anticipated that the separation point will move downstream and the velocity defect of the wake will become small owing to the increase in the velocity near the sphere's surface if fluid slip occurs at the surface for the Reynolds number range considered here.

A vortex loop appears in the wake when the Reynolds number is increased. Fig. **8a** and **b** show photographs of the vortex loop that forms behind sphere. In addition in these figures, we show schematic illustrations of the vortices for clarifying their configuration. A one-way vortex loop begins to form behind a sphere at Re of about 300. However, as Fig. **8a** shows, for the wake of a sphere with the highly water-repellent surface never developed into the Kármán vortex from the twin vortex. The release of the vortex and the vortex street were not observed; however unstable twin vortex became apparent. It is known experimentally that for a smooth surface the vortex loop mode is regular for Reynolds numbers of 300 ~ 420 and irregular for Reynolds numbers of 420 ~ 800. Fig. **8b** shows the vortex loops of smooth surface spheres and sphere with the highly water-repellent surfaces, obtained at Re=430. The Vortex loop of the smooth surface sphere, shown in Fig. **8b**, has greater stability than the waving Vortex tube of the sphere with highly water -repellent surface shown in Fig. **8b**.

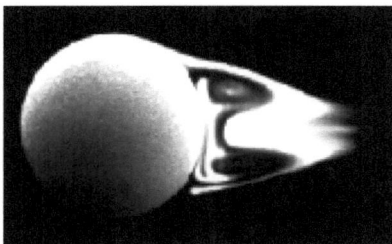

(a) Normal smooth surface sphere   (b) Sphere with highly water-repellent surface

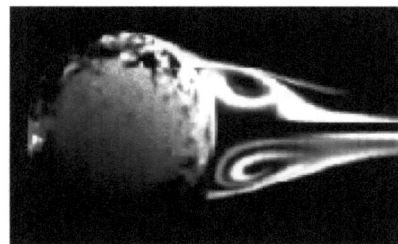

Figure 7: Vortex ring at Re = 100.

(a) Smooth surface sphere

(b) Sphere with highly water-repellent surface

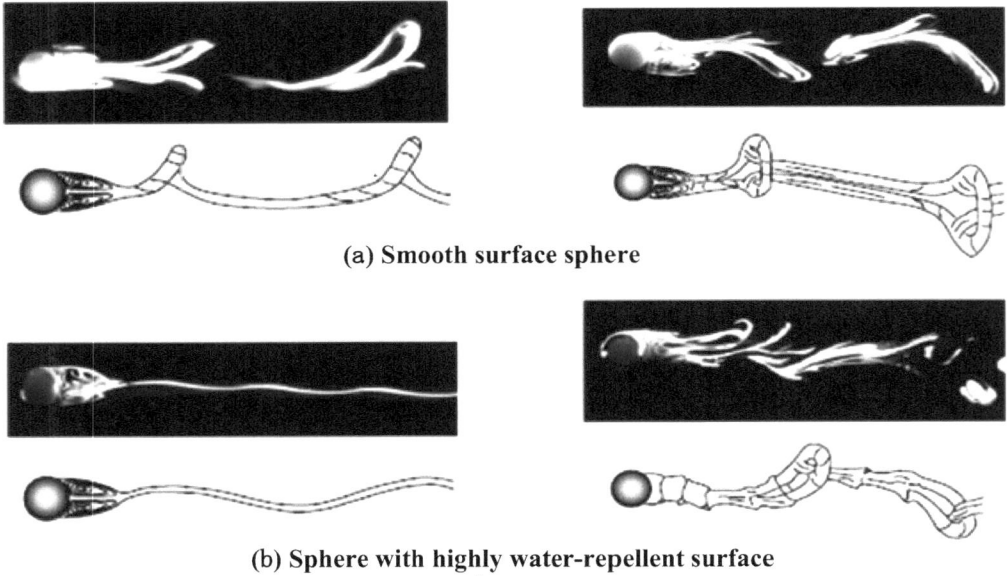

Figure 8: Vortex loop.

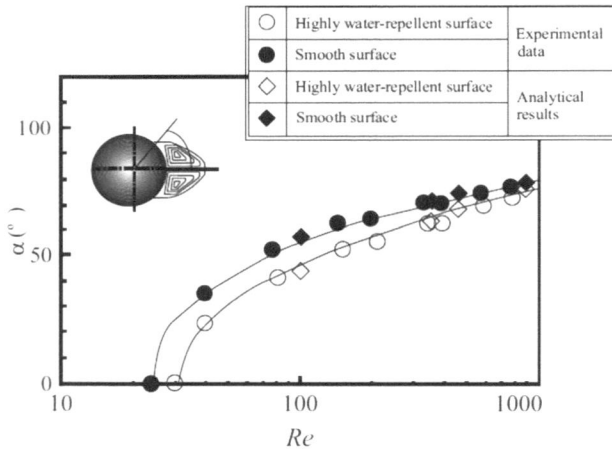

Figure 9: Separation angle *versus* Reynolds number.

It is important to determine separation points for clarifying the characteristics of a boundary layer around a sphere. The separation points were determined by flow visualization in the Reynolds number range of 25 to 850. Fig. **9** shows the separation points obtained experimentally from visualization photographs. The separation angles for the smooth surface sphere agree well with the experimental results reported by Taneda [9]. However, the angles are smaller by $5°$ to $10°$ for the sphere with the highly water-repellent surface. In other words, the separation points move downstream.

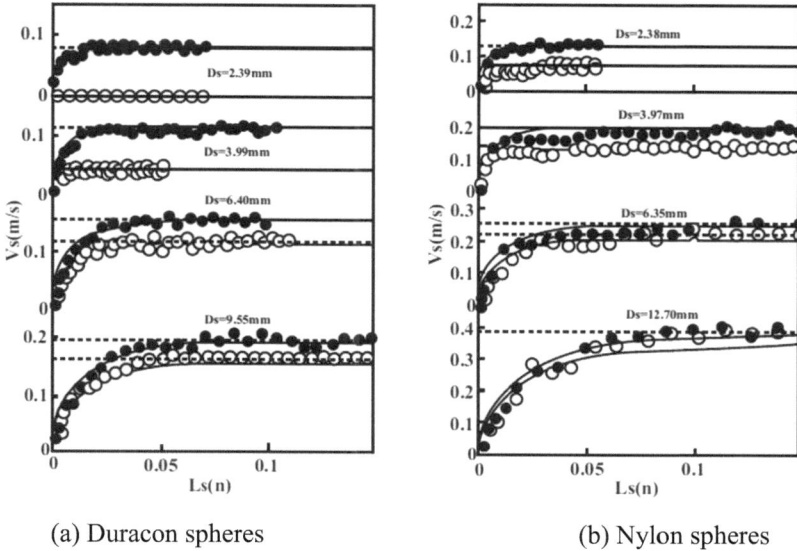

(a) Duracon spheres   (b) Nylon spheres

Figure 10: Fall velocities.

Fig. **10** shows the experimentally measured fall velocities. In this figure, the solid lines indicate the numerical results obtained using Eq. (8-1). The calculation was performed for a sphere that was falling freely though a Newtonian fluid. In Eq. (8-1), we assumed K= 0.5, and Eqs. (8-2) ~ (8-5) were substituted into Eq. (8-1) for determining the drag coefficient. For a sphere with the highly water-repellent surface, we calculate plus the gas membrane in Eq. (8-1). As described above, the gas membrane thickness values were 61μm and 91μm for the Nylon and Duracon spheres, respectively, and the thickness remained nearly unaffected by the sphere diameter. The dotted lines in Fig. **10** indicate the terminal velocity. Experimental data on the terminal velocities shows that the terminal velocity of a sphere with the highly water-repellent surface is lower than that of a smooth surface sphere. The shrinkage is larger for smaller diameter. Consequently, it is seen that for the sphere with the highly water-repellent surface, the experimental data yields higher velocity values than the calculated results for the smooth surface spheres, the experimental results are in good agreement with the results obtained from Eq. (8-1). The reduction in drag is attributed the gas membrane thickness.

Fig. **11** shows the experimental drag coefficients calculated using the measured terminal velocities obtained from Fig. **10**. The solid line in the figure denotes comparison to the Lapple-Shepherd's line [10] for a smooth surface sphere. It was the most large error that the reported drag coefficient measurement at Re=7.2 is best estimate of the result, and with 99% confidence, true value is believed to lie within 1.95% of the estimated result. The experimental data for the sphere with the highly water-repellent surface lie below the solid line for the low-Re range. Thus drag reduction occurs in this range. Fig. **12** shows the drag reduction ratio. The drag reduction ratio is defined as follows

$$RD = \frac{\left(C_{ds} - C_{dh}\right)}{C_{ds}} \times 100 \left(\%\right) \tag{8-6}$$

where, C_{ds} and C_{dh} are the experimentally obtained drag coefficients of sphere with the smooth and the highly water-repellent surface, respectively. The drag redaction ratio increases with a decrease in Re. The maximum drag reduction ratio obtained in this study was 28.5% at $Re = 7.2$ although the peak of the maximum value could not obtained experimentally. The drag reduction phenomenon vanished at Re $\cong 10^4$.

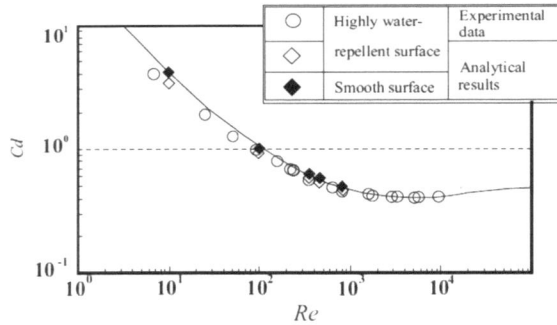

Figure 11: Drag coefficients.

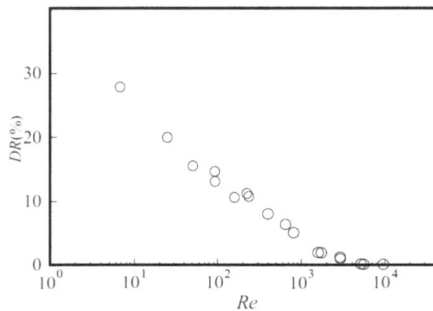

Figure 12: Drag reduction ratio.

4. ANALYTICAL APPROACH

Analysis of the experimental data clearly showed that the behaviors of the separation points and the wake change are induced by the existence of a gas liquid interface on sphere with the highly water-repellent surface and that drag reduction occurs at Re $\leq 10^4$. In this section, we present the analytical results obtained by applying the fundamental equation and the effective slip boundary condition for the apparent fluid slip of the highly water-repellent surface.

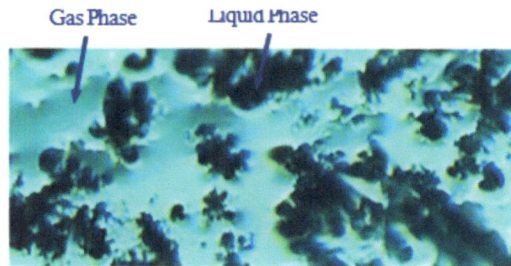

Gas Phase   Liquid Phase

(a) Micrograph of sphere surface in water

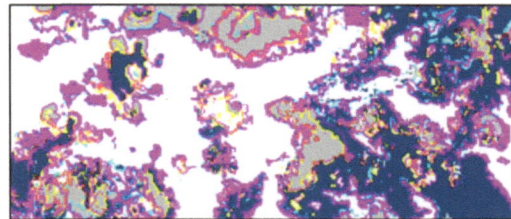

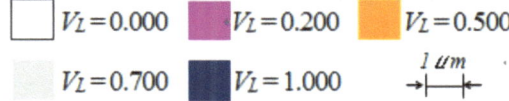

$V_L = 0.000$ $V_L = 0.200$ $V_L = 0.500$

$V_L = 0.700$ $V_L = 1.000$ $1 \mu m$

(b) Image processing of Figure (a)

Figure 13: Photograph of sphere surface in water.

Laminar drag reduction was analyzed in the previous chapter using Navier's hypothesis and Eq. (1-1) as the fluid slip boundary conditions. However, although it is easy to perform the analytical calculations, it is difficult to describe the actual phenomenon of an apparent fluid slip on the highly water-repellent surface. Thus, interface tracking was used in the numerical method for flows with a liquid gas interface. This method can be used for estimating the results of phenomena with changing liquid gas interfaces because the equations for the conservation of mass and momentum are analyzed for local liquid and gas phases; it is difficult to use this method for large-scale flow system calculation.

(a) Gas and liquid in the fractal appearance

(b) Analytical model with surface cell

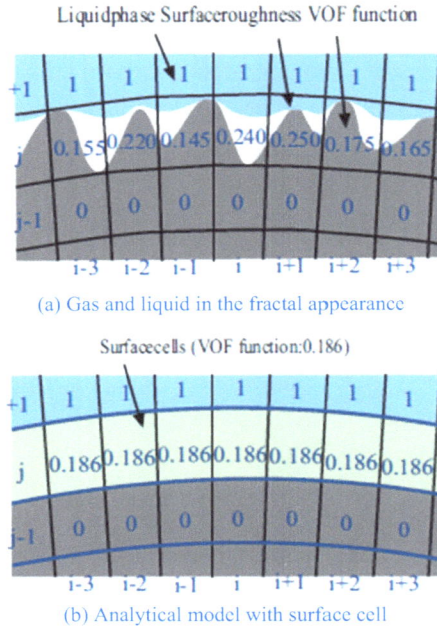

Figure 14: Sphere surface modeling.

A differential space in a two-phase flow with a gas-liquid interface was filled with the gas or liquid phases, respectively. Assuming that the phases do not change, the fundamental equations of incompressible viscous flows give the continuity equation and the Navier-Stokes equation as follows:

$$\nabla \cdot u = 0 \tag{8-7}$$

$$\frac{\partial u}{\partial t} + \left(u \cdot \nabla \right) u = -\frac{1}{\rho} \nabla p + \frac{\mu}{\rho} \Delta u + F_S \tag{8-8}$$

where F_S denotes the volume force of interfacial tension, as in Brackbill *et al.*, [11]. Numerical simulations were performed using the finite volume method. The pressure and velocity values were calculated using the second-order upwinding scheme. The pressure-velocity coupling was determined by using the Pressure-Implicit with Splitting of Operators (PISO) algorithm introduced by Issa [12, 13]. The volume-of-fluid (VOF) method, which is based on the donor-acceptor approach introduced by Hirt and Nichols [14] was used for tracking gas-liquid interface. The *VOF* function *F* is defined to be equal to 1 in the liquid phase and 0 in the gas phase:

$$F = \frac{V_L}{V_L + V_G}$$ (8-9)

where V_L and V_G denote the volumes of the liquid and gas phases, respectively in the differential space,

From Eq. (8-9), it can be seen that the values of F are in the interval [0, 1].

The density of fluid in a differential space, ρ, is expressed as follows:

$$\rho = \rho_G (1 - F) + \rho_L F$$ (8-10)

where ρ_L and ρ_G denote the densities of the liquid and gas phases, respectively.

In contrast, the deformation or transfer of the interface is calculated from the following equation:

$$\frac{\partial F}{\partial t} + \mathbf{u} \cdot \nabla F = 0$$ (8-11)

We used the Continuum Surface Force (CFS) model introduced by Brackbill *et al.* for converting interface surface tension to body force.

The CSF model is given as follows:

$$F_{sv} = \sigma \cdot \kappa \cdot \frac{\nabla \cdot c}{|c|}$$ (8-12)

where, F_{sv} and c denote the body force and the eigenvalue function, respectively. c is given as ρ of Eq. (8-10). σ and κ are the surface tension and the local cuvarture of the interface, respectively.

Thus far, no accurate method has been established for expressing the boundary condition of wettability. The wettability between liquid and solid surfaces can be assumed to influence not only the properties of the liquid and the solid but also the fractal structure of the surface. A micrograph of a hydrophobic surface with a contact angle of 150° sinking through water is shown in Fig. **13a**. Black parts of the figure are parts where water touches the solid surface directly, and white parts of the figure are those where the air is trapped on the solid surface. The volume ratio V of the surface in contact with water is defined as follows:

$$V = \frac{V_L{'}}{V_L{'} + V_G{'}}$$ (8-13)

where $V_L{'}$ and $V_G{'}$ are the volumes of the liquid and gas phases, respectively, near the gas liquid solid interface. Fig. **13b** shows the processed version of the image of Fig. **13a**. The volume ratio V in Eq. (8-13) was found to be 0.186 in the gas liquid solid boundary area by averaging. The volume ratio V is given as a wet boundary condition of the surface. In this analysis, $F = 0.186$ is given as a surface boundary condition for the hydrophobic surface of $V = 0.186$. The wet boundary condition is defined at the center of the cell.

The size of the computational domain in this study for flow past a sphere was 16 mm × 5 mm × 5 mm. The sphere was set at a distance 8 mm from the upper region. The diameter of the sphere was 1 mm on the surface where the hydrophobic effective slip boundary condition was taken into consideration. Rectangular coordinates were used as analytical coordinates. The calculation was performed in three dimensions. The total number of cells was 500,000. Generally, the scale that controls the characteristics of gas liquid two phase flows ranges from 0.1 µm to 10 mm. Thus, an optimum scale for numerical simulation had to be selected.

Because the purpose of this study was to clarify the relationship between fluid slip and wettability of the sphere's surface, as well as to quantify the flow around a sphere, the minimal cell size was set to 10 µm to allow for expression of gas liquid interface.

Fig. **14** shows the typical values of the *VOF* function on the cell surface under wet boundary condition. The *VOF* function of the cell surface was the volume ratio $F = 0.186$. This model was adopted for clarifying the effect of hydrophobic surface fluid slip on the formation of a wake behind a sphere. The analysis used the input condition of the velocity profiles of a laminar uniform flow, as well as the output condition of free outflow (velocity gradient was zero). No-slip surface boundary conditions were used for the surface of the sphere. Initially, the surface of the sphere was covered with a 100 µm thick gas phase film. The liquid phase was water, gas phase was air, and temperatures of both phases were 300 K. The Reynolds numbers used in this analysis were 100, 350, 450 and 650. The calculation was continued for allowing the flow to develop fully. A fixed time step of 1.00×10^{-3} sec was used.

(a) Smooth surface sphere   (b) Sphere with highly water-repellent surface

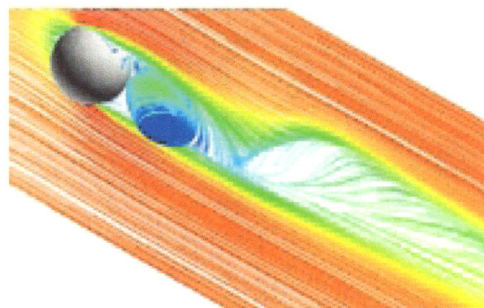

Figure 15: Vortex ring behind sphere at Re = 100.

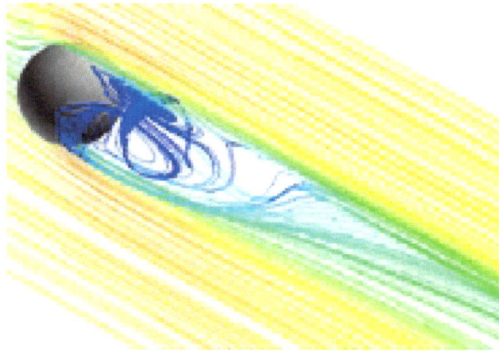

(a) Smooth surface sphere   (b) Sphere with highly water-repellent surface

Figure 16: Vortex loop and vortex ring behind sphere at Re=350.

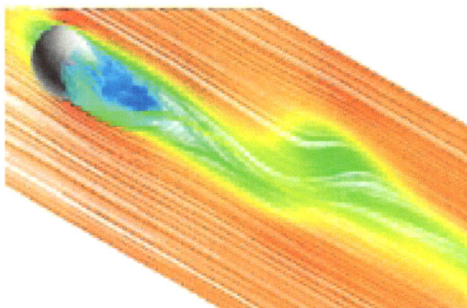

(a) Smooth surface sphere   (b) Sphere with highly water-repellent surface

Figure 17: Vortex loop and vortex tube behind sphere at Re=450.

(a) St=5,000

(b) St=10,000

(a) St=1,000

(b) St=2,000

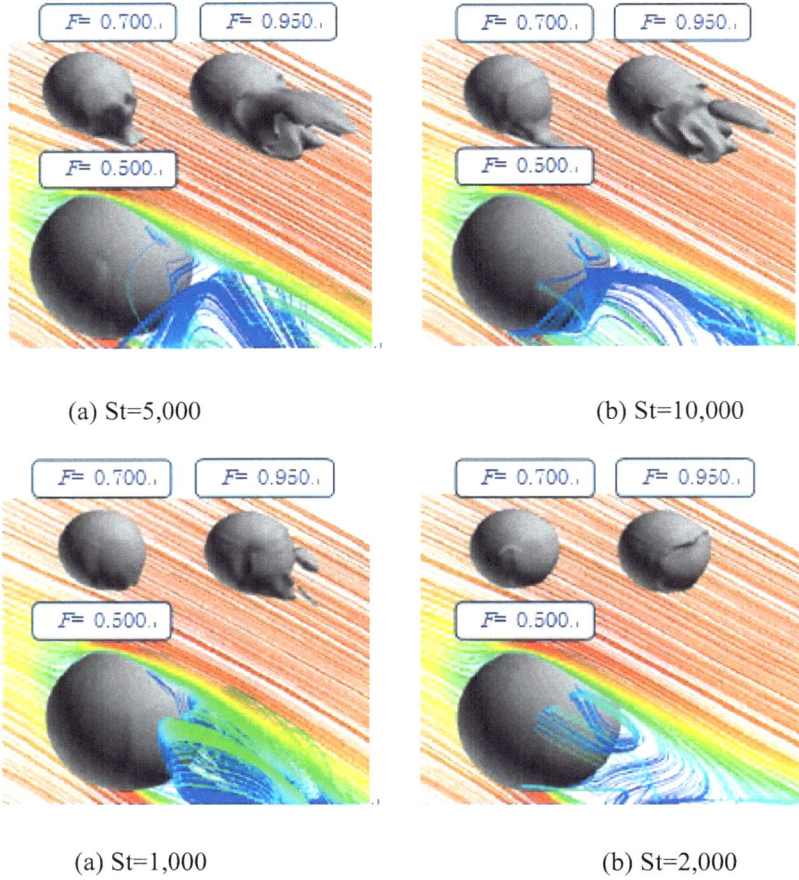

Figure 18: Evolution of gas liquid interfaces for sphere with highly water-repellent surface at Re=450.

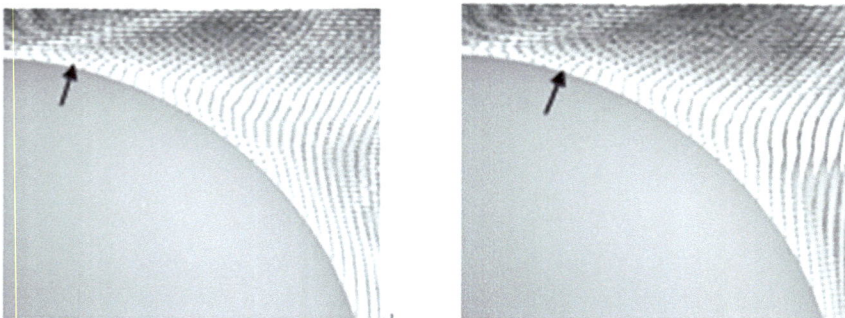

(a) Smooth surface sphere   (b) Sphere with highly water-repellent surface

Figure 19: Numerical results of velocity profiles of around a sphere at Re=450.

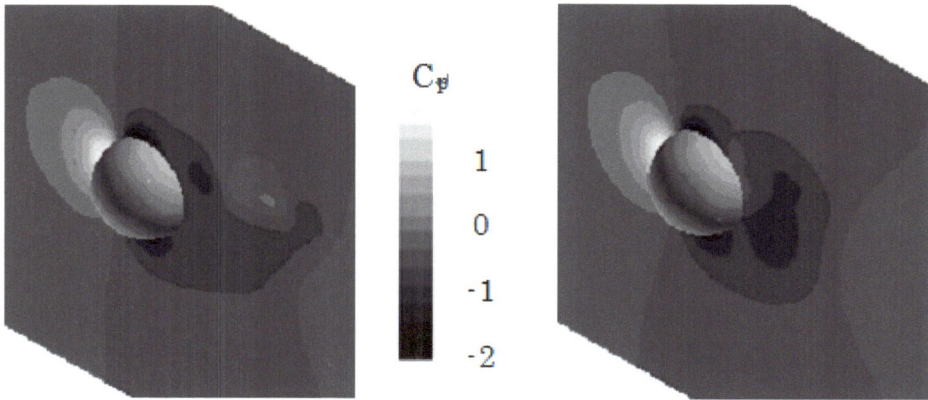

(a) Smooth surface sphere   (b) Sphere with highly water-repellent surface

Figure 20: Numerical results of pressure profiles of around a sphere at Re=250.

Fig. **15a** shows the numerical results obtained for the patterns of a flow past a smooth surface sphere at Re = 100. The figure shows streamlines after 100,000 steps. It is seen that a permanent vortex ring is formed behind the sphere. Fig. **15b** shows the results obtained for a sphere with the highly water-repellent surface, these results were obtained for Re = 100 under the effective slip boundary condition and with the *VOF* function of the surface cells set to a volume ratio of F = 0.186. The figure shows the streamlines after 100,000 steps. It is seen that a vortex ring is formed behind the sphere. The flow visualization shown in Fig. **7** agrees qualitatively with this numerical result. The vortex ring of the sphere with the highly water-repellent surface narrower compared with the analytical results that were obtained for smooth surface spheres (Fig. **15a**). The is ascribed to the occurrence of fluid slip under the influence of the gas phase film near the surface, thus leading to downstream movement of the separation point.

Fig. **16a** shows the numerical result obtained for the patterns of a flow past a smooth surface sphere at Re = 350. The figure shows streamlines after 100,000 steps. It is seen that a one-way vortex loop is formed behind the sphere. The flow visualization results shown in Fig. **8a** agrees well with the numerical results. Fig. **16b** shows the results obtained for a sphere with the highly water-repellent surface. These results were obtained for Re = 350, under the effective slip boundary condition and with the *VOF* function of the surface cells set to a volume ratio of F = 0.186. The figure shows streamlines after 100,000 steps. It is seen that a vibrating vortex ring is formed behind the sphere. These results agree with the flow visualization results shown in Fig. **8b**. The reason for this is that the separation points move downstream, and the velocity loss area decreases.

Fig. **17a** shows the numerical results obtained for flow past a smooth surface sphere at $Re = 450$. The figure shows streamlines after 100,000 steps. It is seen that a rolling vortex loop is formed behind the sphere. The flow visualization results shown in Fig. **8a** agree well with these numerical results. Fig. **17b** shows the case of a sphere with the highly water-repellent surface at $Re = 450$, under the wet boundary condition, and with the *VOF* function of the surface cells set to a volume ratio of $F = 0.186$. The figure shows streamlines after 100,000 steps. It is seen that a wavy vortex tube is formed behind the sphere. The flow visualization results of Fig. **8b** agrees well with the analytical results. The wake of a smooth sphere is also translated to a vortex tube as the Reynolds number increases up to $Re = 800$. Compared with the analytical result for the smooth surface sphere in Fig. **17a**, the waving vortex tube of the sphere with the highly water-repellent surface is less stable.

We carried out a numerical simulation of the gas liquid interface to examine the effect of the time step on the interface. Fig. **18** shows the evolution of the gas liquid interfaces at the highly water-repellent surfaces around the sphere for various numerical time steps at $Re = 450$. The contour figures of the *VOF* functions that are 0.500, 0.700 and 0.950 represent the interfaces. It can be assumed that $F = 0.500$ approximate the gas liquid interface, and $F = 0.950$ approximates the liquid phase of water. It is seen that the gas liquid interfaces on the spherical surface are transformed with time, and the interfaces begin to collapse near the separation point. However, not all gas phases are washed away in the downstream direction. No deformation of the gas liquid interface is observed after 10,000 steps. The interface stabilized when the sphere was covered by gas phase film. In contrast, in the case of a smooth surface sphere, all gas phases on the spherical surface under the initial condition were washed away with time and the sphere was never covered by stable gas liquid interface.

The drag of a sphere was calculated under the effective slip boundary condition assumed in this study. The analytical results are shown in Fig. **11**. The results for a smooth surface sphere with the no-slip boundary condition are shown in Fig. **11** for comparison with the no-slip experimental results. Total drag is given as the sum of viscous and pressure drags, and viscous drag is the predominant region in the total drag within the low Re range. The analytical results show that the drag of a sphere's surface with fluid slip decreases in comparison with that of a smooth surface sphere. The drag reduction is about 10% for Re > 100; it increases with decreasing Reynolds number. The results agree well with the experimental results obtained for a sphere with the highly water-repellent surface.

It was clarified in the visualization experiment that the separation angle of a sphere with highly water repellent surface decreases in comparison with that of a smooth surface sphere for Reynolds number ranging from 20 to 800. Because the separation of flow is closely related to the drag reduction, the velocity profile near the separation point was calculated. The results for Re = 450 are shown in Fig. **19**. For comparison, the results for the smooth surface sphere are shown well. In Fig. **19**, the arrow shows the separation point. It is seen that the separation in the case of a sphere with highly water-repellent surface, the separation point moves downstream about 10 degrees. The calculation results agree well with the experimental results shown in Fig. **9**. The change of flow separation will generate a decrease in the pressure drag.

We calculated the pressure profile in order to grasp the reason for the drag reduction of a sphere with highly water-repellent surface. Fig. **20** shows the calculation results, and these results are illustrated by the pressure coefficient which is defined as follows:

$$C_p = \frac{p_0 - p_\infty}{\rho U^2 / 2} \tag{8-14}$$

where p_0 and p_∞ denote the pressure at the surface and in the region of uniform velocity, respectively. The parameters ρ and U are the fluid density and the uniform flow velocity, respectively. The integral value of the pressure profile of a sphere with highly water-repellent surface is smaller than that of a smooth surface sphere. Thus, the pressure drag is smaller for the case of a sphere with highly water-repellent surface. The vortex tube is generated at Re=450 in the case of a sphere with highly water-repellent, and the pattern of flow for the smooth surface sphere differ from that obtained for the sphere with highly water-repellent surface. It is seen that the decreased separation angle affects the pressure profile; however, the relationship between the vortex tube and the profile is not grasped in detail at present.

SUMMARY

In this chapter, we took the first step towards developing a novel insightful approach for the quantitative determination of the role of the gas liquid interface at highly water-repellent surface in millimeter scale flow system, because, although being simple, the analytical method that was presented in the preceding chapter is not based upon the actual phenomena for this fluid slip. In addition, a model of the wet boundary condition for apparent fluid slip was proposed, which allows calculating the pattern of a flow past a sphere and the total drag in flow systems. Flow visualization results that were presented here showed that no vortex loop existed at *Re*<400 for a sphere with a highly water-repellent surface, and a vortex tube began to form behind the sphere at *Re*<400. Compared with the results obtained for a smooth surface sphere, the waving vortex tube of a sphere with highly water-repellent surface is less stable. The separation points moved downstream compared with the separation points obtained for a smooth surface sphere, because a gas-liquid interface existed at the surface. The evolution of gas liquid interfaces at highly water-repellent spherical surfaces was clarified, and it was shown that such interfaces are transformed with time and start collapsing near the separation points. For the interpretation of results, it will be necessary to consider the mechanism of apparent fluid slip. Drag reduction occurred in the flow and the maximum drag reduction ratio was 28.5 % at *Re* = 7.2.

A flow with fluid slip was modeled by observing the gas liquid interface at highly water-repellent surface in millimeter-scale flow systems. Moreover, a flow around a sphere with a highly water-repellent surafce was analyzed by applying the model and basic equations. The analytical results obtained by this method agreed well with the experimental results of the flow pattern and the drag coefficient for Reynolds number ranging from 5 to 400. The results suggest that the wet boundary condition for fluid slip proposed in this work is valid for a highly water-repellent surface.

REFERENCES

[1]   Watanabe, K. & Fujita, T. (2003), "Motion of a Highly Water-Repellent Sphere in Newtonian Fluids", *Proc. ASME/JSME 4th Joint Fluids Summer Eng. Conf.*, FEDSM2003-45787, pp. 2513-2518.
[2]   Fujita, T & Watanabe, K., (2006), "Numerical Simulation of Flow Past a Macroscopic Sphere with a Hydrophobic Surface", *Trans. of JAME*, Ser. B, **72**, 714, pp.83-90.
[3]   Fujita, T & Watanabe, K., (2006), "Flow Visualization Past a Hydrophobic Surface Sphere with a Gas-Liquid Interface", *Trans. of JAME*, Ser. B, **72**, 714, pp.91-97.
[4]   Fujita, T & Watanabe, K., (2004), "Gas-Liquid Interfaces and Flow Characteristics of a Highly water-repellent Sphere in Newtonian Fluids", *Trans. of JAME*, Ser. B, **70**, 692, pp.146-152.
[5]   Stokes, G.G., (1844), "On the theories of internal friction of fluids in motion and of the equilibrium and motion of elastic solid", Trans. Cambr. Phil. Soc., **8** (9), pp.287-319.
[6]   Schillar, L., & Nauman, (1933), "Uber die grundlegende Berechnung bei der Schwekraftaufbereitung, Ver. Deutch Ing. **77**, pp. 318-320.
[7]   Carey, W. W., (1970), "Settling of Spheres in Newtonian and Non-Newtonian Fluids", Ph. D., Thesis, Syacuse Univ., New York
[8]   Brauer, H. (1973),"Impuls-, Stoff- und Wärmetransport druch die Grenzfläche kugelförmiger Partikeln",*Cremie-. Ing. Techn.*, **45**. pp.1019-1103.
[9]   Taneda, S. 1956, "Studies on Wake Vortices (Ⅲ): Experimental Investigation of the Wake behind a Sphere at Low Reynolds Numbers", *Rep. Research Inst. for Applied Mech. Kyushu Univ.*, **Ⅳ**, 16, Oct.

[10]   Lapple, C. E. & Shepherd, C. B., (1940), "Calculation of particle trajectories", *Indust. Eng. Chem.*, **32**, 5, pp.605-617.

[11]   Brackbill, J. U., Kothe, D.,B. & Zemach, C., (1992), "A Continuum Method for Modeling Surface Tension", *J. Comp. Phys.*, **100**, pp. 335-354.

[12]   Issa, R. I., (1985), "Solution of the Implicitly Discredited Fluid Flow Equations by Operator Splitting", *J. Comp. Phys.*, **62**, pp. 40-65.

[13]   Issa, R. I., (1991), "Solution of the Implicitly Discredited Reacting Flow Equations by Operator-Splitting", *J. Comp. Phys.*, **93**, pp. 388-410.

[14]   Hirt, C. W. & Nichols, B. D., (1981), "Volume of Fluid (VOF) Method for the Dynamics of Free Boundaries", *J. Comp. Phys.*, **39**, pp. 201-225.

Subject Index

Subject Index

C

Circular cylinder 71
Contact angle 9
Couette flow
    Flow between two coaxial rotating cylinder 43
    Enclosed rotating disk 57
CSF model 95
Circular pipe flow
    Newtonian fluids 11
    Non-Newtonian fluids 21
Drag coefficient
    Circular cylinder 77
    Sphere 87

D

Drag reducing additives 5
Drag reducing wall 5, 35
Drag reduction ratio
    Rotating disk 66
    Sphere 92
Duct 29

F

Flow angle 60
Flow patterns
    Rotating disk 59
    Circular cylinder 74, 81
    Sphere 89, 90, 97, 98
Friction factor
    Newtonian fluid 14, 26, 33, 38
    High molecular polymer solutions 22
    Surfactant solutions 24

G

Göltler vortex 48

H

Hagen-Poiseuille flow 12

M

Moment coefficient
    Two coaxial rotating cylinders 45
    Rotating disk 57, 65, 68

N

Navier-Stokes Equation 94
Navier's hypothesis 5, 18
Non-Newtonian fluid 21, 45

P

PIV 41
Power law model 22, 45
Power law Reynolds number
    Circular pipe flow 22
    Flow between two coaxial rotating cylinders 45
Pressure coefficient 101
Pressure gradient 13

R

Rotating disk 55
Riblets 4

S

Separation
    Circular cylinder 74
    Sphere 90, 98
Sphere 83
Sliding constant 6
Slip velocity 5, 18, 48
Strouhal number 77
Surfactant solutions 24

T

Taylor cell 52
Taylor vortex 42, 49
Turbulent drag reduction 3, 22
Twin-vortices 74, 81
Two coaxial rotating cylinders 41

V

Velocity defect 76
Velocity profile
    Pipe flow 18
    Duct flow 30, 33, 37
    Two coaxial rotating cylinders 48
    Enclosed rotating disk 61
von Kármán
    Momentum integral equation 68
    Vortex 74, 90

W

Wake 78
Wall roughness 9, 31, 35, 86, 94
Wall shear stress 18, 47

www.ingramcontent.com/pod-product-compliance
Lightning Source LLC
Chambersburg PA
CBHW041720210326
41598CB00007B/716